Dancing Dads, Defective Peeps & Buckeye Misadventures

Dancing Dads, Defective Peeps & Buckeye Misadventures

by Joe Blundo

ORANGE FRAZER *PRESS*
Wilmington, Ohio

ISBN 1-882203-43-7
Copyright 2004 by The Columbus Dispatch

No part of this publication may be reproduced in any material form (including photocopying or storing in any medium by electronic means and whether or not transiently or incidentally to some other use of this publication) without the written permission of the copyright holder except in accordance with the provisions of the Copyright, Designs and Patents Act 1988.

Additional copies of *Dancing Dads, Defective Peeps & Buckeye Misadventures* may be ordered directly from:

Orange Frazer Press
P.O. Box 214
Wilmington OH 45177

Author photograph: Neal Lauron, *The Columbus Dispatch*
Cover illustration: Andrea Levy
Inside illustration: Margaret Morgan
Book and cover design: Tim Fauley, Orange Frazer Press

Telephone 1.800.852.9332 for price and shipping information
Website: www.orangefrazer.com

Library of Congress Control Number: 2004105772

To my wife, Deborah, and my children, Noah and Celia

Acknowledgements

THIS BOOK WOULDN'T EXIST were it not for so many people at *The Columbus Dispatch* who gave me opportunities over the years. I'm grateful to all of them. In particular, I want to thank Mike Curtin, who made me a columnist in 1997, and Mary Lynn Plageman, who pushed to make this book possible and provided valuable editing.

Table of Contents

Adventures

Modern Bewilderments

Introduction

AFTER TWO DECADES IN THE NEWSPAPER BUSINESS, I somehow ended up with a job that allows me to be paid for swimming badly, overeating at the state fair and putting words into the mouths of dinosaurs.

I try not to think about the fortunate combination of circumstances that took me from writing about real estate to writing about Heidi Klum's underwear. To contemplate it too deeply would be to jinx the whole, happy situation, I fear.

Writing my newspaper column, So To Speak, has many rewards, but perhaps the most satisfying is that you learn once and for all that you're not crazy. Your offbeat observations and bizarre twists of thought actually make enough sense to warrant publication. People can follow your train of thought!

What's more, column-writing makes everyday frustrations tolerable. The guy who brings 27 items to the 12-items-or-less line at the grocery store simultaneously outrages and delights me: He may be annoying, but at least I can get a column out of him.

On top of all that, I get to write the column in Columbus, Ohio, an endearingly funny place. When I moved here in the 1970s, I was amazed at the strange folkways: the obsession with Ohio State football, the agricultural exhibitionism of the Ohio State Fair, the love of mulch.

Over the years, the town has just grown more entertaining. It frets about its body-mass index. It panics over the weather. Every few years, it fires a coach as a form of group therapy.

I love the place. It's big and friendly, ever insecure, always bursting with amusing developments.

If I didn't have Columbus, though, I'd still have my family. When there's a columnist in the house, everyone else must expect to be pressed into service as a literary device. Every transition — from high-school graduation to the delivery of a new dryer — inspires a column. My wife and kids have been surprisingly patient about this.

Occasionally, I do write serious columns. I chose not to include any of them in this book. The nation is overrun with commentators writing importantly about events of the day. Turn to them for cogent explanations of the state of the world. Most of the time, I find the world largely inexplicable.

As the dates in this book show, I began writing columns long before I became a columnist. Someone finally noticed, and I was given the title in 1997. Since then, I've churned out about a thousand. I thought it would be easy to choose 60 or so for this book, but like most other writers I have an inflated opinion of my own work. So I agonized for days, as if it mattered deeply to the world of literature whether I included the column about the time I made Marshmallow Peeps.

Anyway, here are my choices, Peeps and all. I hope you find them amusing.

Joe Blundo
Columbus, 2004

One

Family Matters

Somehow, dads just aren't what they used to be

WHAT'S A DAD DO?

Had you asked me that question at age 8 or 9, I probably would have said that dads go to work, play cards, smoke, golf, shave, drink the occasional beer, wear hats and grill hot dogs when the need arises.

When I was a kid, dads were readily identifiable. They were the big people who didn't run. Running simply wasn't a thing dads did.

Nowadays, you go to a park and there are dads running all over the place. The ones who aren't running are dressed as if they intend to at any moment. Last summer, because I was dressed for it, I played tag at a park and twisted an ankle in full sprint.

My dad would never have dressed himself in such a dangerous way — he wore dress shoes to picnics. So did the other picnicking dads of my acquaintance.

Occasionally, back then, you could coax a dad into running. But what they called running was more like an animated walk that jingled the change in their pockets. Often, they'd run with a cigarette to reinforce the point that this wasn't something they expected to be doing for long.

I don't know why dads didn't run — maybe they thought World War II had been enough exercise.

Dads watched sports, but the only one I knew them to play was golf. Not this dreadful business golf that people play now to impress one another. The dads I knew were mostly blue-collar workers. The idea of golfing as a condition of employment would have struck them as absurd.

Dads went out after work or on a Saturday morning and golfed for fun. They never talked about their game after it was over. They came back sunburned, a couple of beers under their belts, and went to sleep.

Dads didn't drink pop, on the golf course or anywhere else. And they certainly wouldn't drink it for breakfast, the way some dads do now.

My guess is that the dads of the old days would have been more suspicious of a guy belting down Diet Coke at breakfast than of one swilling whiskey. This has stuck with me, a breakfast coffee drinker. To this day, I have a hard time taking a morning pop drinker seriously, even if he has a car phone.

I always thought that when I became a dad I would wear a hat. The standard-issue dad wore a hat when I was a kid. Apparently, it had nothing to do with the weather — I remember only those vaguely gangsterish dress hats that couldn't possibly have offered any warmth and were prone to blow away in a light breeze.

Evidently, the point was to be seen in a hat, not to keep warm. A dad in a ski cap? Unheard of.

I figured the kids of my generation would grow up to be dads, and most of us did. I didn't know we would take all our toys with us. We play board games, for heaven's sake.

I can't imagine the dads of my childhood getting together over a game of Pictionary. They would have played cards, something complicated like pinochle.

We would hang around the table when dads played cards, to soak up the atmosphere. There was no possibility of being invited to play, and it wasn't a great idea to ask too many questions.

There were hidden dad dimensions back then. One of them was cooking. Dads were special-event chefs — they took over when there was food to be grilled or a turkey to be carved.

But most amazing to me were the dancing dads. They all had the ability. We would go to a wedding reception and, all of a sudden, these normally sedentary, card-playing men would get up, grab their wives and reveal that they knew how to dance.

I figured I would grow up and acquire this skill, too, but here I am, 13 years a dad, with no dancing ability. I dance the way dads used to run — in short intervals and with ample evidence that this is something I don't often do.

"He can run but he can't dance?" That's what the incredulous dads of my era would have said. Then they would have gone to play cards, still chuckling about how that boy better grow up.

—Saturday, June 14, 1997

Veteran parent offers a few tips for beginners

THIS SUMMER I WILL MARK MY 15th anniversary of parenthood, which is another way of saying that my son is turning 15.

He and his 7-year-old sister have taught me many things about raising kids.

I would like to pass on a few bits of parental wisdom now, before either one gets a driver's license and becomes capable of doing something that might cast serious doubt on my expertise.

Here are some lessons for rookie parents:

• Kids will forget everything you say but never fail to notice anything you do. For 18 years, every move you make will have the potential to influence them. Try not to be self-conscious about this.

• The biggest battle you will fight in the first 10 years will be over food.

Kids advance from throwing food to fingerpainting with it to insisting that it be cut a certain way to shunning food of one color if it has touched food of another color. For years, they will become nauseated at the very mention of say, asparagus, until they abruptly and inexplicably begin refusing to eat anything but asparagus.

They will go through periods when they can subsist for an entire week on three green beans and a glass of water, and other periods when letting them eat directly from the meat case at Kroger wouldn't satisfy their hunger.

You can avoid food battles by serving pizza at every meal for an entire decade.

• The second biggest battle you will fight in the first 10 years will be over dressing for the weather. Look at it from their perspective: How would you dress if you didn't care what the neighbors thought and you knew that there was always someone hovering nearby with an emergency sweater?

• Speaking of clothes, never dress a kid too warmly if you are taking her to a playground on a cold winter day. She will comfortably play for hours while you slowly succumb to hypothermia.

• People with wacky political agendas will try to convince you that there is no difference between the sexes, save for plumbing. Boys and girls, however, don't read wacky political agendas. They are fundamentally different.

• If you're lucky, your kids will reserve their worst behavior for home, where they feel safe acting on impulses they have learned to suppress in public. If you're not lucky, they will feel safe acting on their impulses everywhere.

• Never buy the first Happy Meal. From then on, the child will be satisfied with nothing less and you will end up with 3,000 cheap Disney action figures.

• Your kids can save you a load of worry if they prove to be good at picking their friends.

• Kids tend to understate illness and overstate injury. If a kid says he's feeling a little tired, it means he has a raging fever and is about to projectile-vomit on everyone within two city blocks. If he runs into your arms screaming, his face contorted into a mask of pain, it means he has a mosquito bite.

• Ignore every prediction other people make about your kids. People love to predict the adult size, intelligence, occupation and temperament of babies. They are always wrong.

• Kids don't like surprises. Always warn them what happens next.

• Siblings love to fight with each other. Experts will advise you not to referee these fights. You, however, are hard-wired to respond to the sounds of kids in distress, even if it's the kind of distress they enjoy. Happy refereeing.

• Kids always wait until they're on a slow, crowded elevator to ask you embarrassing questions about sex, race or personal hygiene. When they do, pretend you're only the baby-sitter and say: "That's a good question to ask your parents. And stop calling me Dad."

• If a child is driving you crazy, wait six months until the next stage of development arrives. That either will solve the problem or replace it with a new one.

• Learn the stages of development. No information is more valuable to a parent.

• Adding a second child to the family is like going up one full point on the Richter scale — the magnitude of impact is 10 times as great. That means 10 times the aggravation and 10 times the joy. So it's a good deal.

—Saturday, June 13, 1998

Some advice for drivers who are younger than ever

MOST OF THE TEEN-AGERS I KNOW are pretty good people, considering the bizarre world adults have created for them to live in.

But are we sure we want them driving at 15½? They're still seeing pediatricians at that age, for heaven's sake. Shouldn't there be some kind of a transition period between bicycles and Land Rovers? The first crop of 15½-year-olds are getting their learner's permits under a new state law that allows them behind the wheel at that tender age when accompanied by a parent or guardian. My son, who is 14 and 11/12ths, will be only too happy to comply with this law in a few short months, I'm sure.

The idea behind the law is to give teens a full six months of instruction before they can venture out solo. But even at a well-schooled 16, they're going to have to contend with roads full of crazy adults, many of whom couldn't be helped with six years of instruction.

Oh, well. There's no holding them back now, so here are some observations about life on the road for these fledglings. Here you go, kids:

• Someday, perhaps in your lifetime, a modern, sophisticated means of transportation will be devised.

In the meantime, we are stuck with this primitive system of individual sheet-metal boxes that burn liquefied dinosaur remains while speeding over narrow concrete slabs. It's a crude, crowded, dirty and dangerous way to travel. Keep this in mind, and you won't get road rage. Road rage is the affliction of deluded people who think car travel is supposed to be carefree.

• There are always a few adults who want to argue that wearing seat belts infringes on their liberty. If their argument impresses you, you're not mature enough to drive.

• A car is expensive. To make it more affordable, plan on keeping it through the administrations of at least three American presidents. Barring Whitewater impeachment hearings, my 1989 Toyota, bought in the waning days of the Reagan administration, will last me until Bill Clinton leaves office in 2001.

Note: This plan has drawbacks. Had I bought a car when Herbert Hoover took office in 1929, I couldn't have replaced it until Harry Truman said goodbye in 1953. I got around this situation by not being born until 1954.

• Buy station wagons. They are the geekiest vehicles available. You will never have to worry whether someone loves you just for your car if you're driving a station wagon. No one will steal it, no one will challenge you to race it, and you will never be tempted to take a corner at 45 mph. At worst, you will get the occasional request to move furniture.

• When stuck in traffic, visualize everyone as driving kiddie cars at an amusement park. This will stop you from taking the situation too seriously. When not stuck in traffic, visualize everyone as carrying cargo that will explode in a mammoth fireball on impact. This will help you keep the proper distance from other vehicles.

• Closely observe the driving habits of trusted adults, and don't imitate any of them. Don't drink and drive, eat and drive, phone and drive, read and drive, apply makeup and drive, play Monopoly and drive, sleep and drive, argue and drive, change diapers and drive or kiss and drive. In fact, if at 16 you're doing anything other than breathing and driving, pull over.

• Adults judge people by their cars. Apparently, they've never heard of modern leases that allow cooks at Taco Bell to drive Lincoln Navigators. Don't judge people by their cars. Judge them by how they drive their cars. Nothing reveals a jerk faster than turning the ignition key.

• The proper ratio for turn signal-to-horn use is 500:1. Many adults have this backward.

• There is nothing more boring than car talk, save for computer talk and anything anyone says on sports call-in shows. Engine displacement, number of valves, type of camshaft (whatever that is) — no one wants to hear you talk about these things.

The only car fact worth mentioning in public is this one: When a car crashes at 60 mph, it stops, but you keep moving at 60 mph until your body hits something hard or sharp. Many adults drive as if they've never heard this, so repeat it often.

• If at all possible, take the bus.

—Saturday, July 11, 1998

He's learned a thing or two in 25 years of wedded bliss

I'M MARRIED. Boy, am I married.

Today is my 25th wedding anniversary. You can't get much more married than that.

I'm so married that people who don't even know me know I'm married. Salesclerks see me eyeing the merchandise and say, "Don't you think you'd better check with your wife first?"

Well, they're right. Consulting with your spouse is about 90 percent of marriage. If you want to get in trouble in a marriage, fail to seek mutual consent. It's a killer.

Most marital arguments stem from failure to consult. I think my parents had a fight once over green peppers. I'm sure it wasn't the peppers themselves but the fact that one had failed to consult the other about the peppers. Those of you who are married know that it's always best to consult, even about vegetables.

If you're single, don't let these revelations scare you. It's not like you have to consult on everything. There is a special non-consulting category called That's Just the Way He/She Is.

Items in this category require no consultation because it wouldn't do any good.

For example, I have this thing about saving tiny amounts of leftover food, then throwing them away three weeks later. Why don't I stop? Many explanations could be advanced, but the best one is That's Just the Way He Is.

A healthy marriage has a lot of compromise, with a little bit of That's Just the Way He/She Is to allow for quirks.

But not too much.

If you go over 10 percent, you can cause trouble in crucial areas, such as your joint credit rating. If the electric bill is six months overdue, That's Just The Way He/She Is won't cut it as an explanation.

My marriage is at 5.8 percent, well within normal limits. It had crept up to 6 percent, but I threw out some old potato salad the other day.

Our marriage has had two eras: Before Kids and During Kids. The third era, Kids Away But Requiring Massive Amounts of Money, is looming.

I can't offer many comments about the Before Kids era because I can't remember. It lasted eight years, so you'd think it would have left stronger impressions. But having kids causes a brain-chemical shift that results in all memories of life before kids being dimmed.

This is nature's way of ensuring that you'll continue to propagate the species even if your pre-propagation life had more sleep and fewer spills.

The During Kids era has had profound effects on our marriage. This is when we began doing all those things we'd just finished rebelling against: going to church, clipping coupons, associating with known Republicans.

It was then that I knew why the most irrefutable piece of advice any parent ever can give is, "Just wait."

I'm not surprised that we're still married after 25 years. I am surprised that I could be married for 25 years and still not feel like a full-fledged adult.

When my parents had been married 25 years, they were instantly recognizable as adults. They'd lived through World War II. They knew how to dance. They wore hats. All the signs were there.

I, a baby boomer raised in more frivolous times, just don't have the same weightiness. It worries me now and then. I want grandchildren to climb in my lap and beg me to tell them brave tales from the old days. What am I going to say? I survived disco?

Oh, well, I can always make something up.

When we got married, our parents told us they hoped we would be as happy as they had been.

Of course, we had no idea what they were talking about. We thought we did, probably, but we were too young and dumb. We couldn't understand, much less cherish, the intricate tangle of connections that grow between two people through many years.

Like all people, we wed on nothing but a promise. Compared with today, we were hardly even married when we got married. But, wow, are we married now.

—Saturday, July 29, 2000

Believe it or not, Dad has learned a few things

ON MONDAY, I WILL HAVE A child who is an adult.

I'm speaking technically, of course. Parents do not recognize their children's adult status until the kids reach 40 or 50, if then. My mother still tells me when to put on a sweater.

Nevertheless, my son's 18th birthday looms as a milestone. It makes me feel older than any birthday I've ever had myself.

On such a noteworthy occasion, I feel compelled to offer advice on the finer points of being an adult.

It would be better if a fully qualified adult were available to do this. I exhibit signs of adulthood — mortgage, voter registration, some graying at the temples — but turn your back on me and I'm liable to play with Legos or something. I don't even golf.

Then again, it doesn't matter. This is a low-risk exercise because 18-year-olds are genetically programmed to be impervious to most of what their parents tell them. The wisdom of what I am about to say, assuming there is any, will not become apparent to him until he has already learned most of it himself.

Despite that, I will forge ahead. Here's what I want to tell the new adult:

• Most people overcook spaghetti. Try to avoid that.

• People will continue to act the way people always have acted. Keep this in mind when a technological development is hailed as a revolutionary breakthrough that will change humanity. More likely, it will simply allow people to do what they've always done, only faster.

If you want a good laugh, go back and read what visionaries were writing about the Internet 10 years ago. It was going to change everything, from shopping to literature. A decade later, it's dominated by porn sites and places where people can auction trinkets for 50 percent more than they're worth.

• It's always cheaper to keep the old car.

• Try going to movies that don't have movie stars in them. They're almost always better.

• No one takes the Bible literally. Or the Koran. Or any other sacred text. No matter who is holding forth on the Holy Book, you are listening to someone's selective interpretation of it. Keep that in mind when you run into people who claim to have found all the answers.

That said, you should never stop looking for the answers.

• Successful conspiracies involving more than two people and lasting more than two days are extremely rare. If someone's argument hinges on a conspiracy theory, the argument isn't worth considering.

• Even if you're reluctant, say yes sometimes when people ask you to volunteer for something. You might be surprised at what you learn.

• When you buy a house, buy one with a front porch.

• Read the books people want to ban. They're usually pretty good.

• If you want people to follow orders, write down what you want them to do. Then tell them. Then tell them again. It's amazingly difficult to get something across to a group.

• When you listen to a diehard liberal arguing with a diehard conservative, you are listening to two people who are equally narrow-minded. Political arguments are largely one set of prejudices jousting with another for theatrical effect. Don't take them too seriously.

• Nevertheless, vote.

• Don't take the news media too seriously, either. At best, we skim a few interesting events off the top of the cauldron every day. The real story is always infinitely more complicated.

• Nevertheless, read.

• However much time you spend watching television in your life will be too much time.

• People often reveal themselves by how they drive. Also, by how they treat servers in restaurants. Remember that.

• If you dislike the job, the money it pays won't be enough, no matter how much it is.

• Get married. Not now, for goodness sake. But someday. Later, have children. Remain married thereafter. Regardless of societal trends, this is still the best way to do things.

—Saturday, August 4, 2001

Dad, An Owner's Manual: Dad requires some special care, feeding

CONGRATULATIONS. YOU ARE THE PROUD OWNER of a dad. Treat him well, and he will give you many years of service.

Before operating your dad, read this manual carefully.

Warning:

• Do not attempt to operate your dad when drowsy. Let the poor guy sleep.

• Shock hazard! If you've been doing some of the same things your dad did at your age, don't tell him. He will be shocked.

• Should you see dense clouds of smoke and bright flames leaping 20 feet into the air, get a plate immediately. Your dad is grilling something.

• If your dad will be performing strenuous tasks (anything he hasn't done in 10 years but insists he still can do), the wearing of eye, ear, nose, throat, head, elbow, wrist, hip, knee, ankle, and foot protection is recommended — for your dad, not you. You should dress lightly so you can run for help.

Section A: Maintenance

• Your dad is a sophisticated, finely tuned instrument. Feel free to tell him that, by the way.

• For optimum performance, feed your dad regularly. Dads prefer a healthful, low-fat diet until they get hungry. Then they'll eat anything.

• After operating your dad for extended periods in hot weather, return him to a horizontal position in a shady location, provide cold beverages and leave him alone for a minimum of two hours.

• Your dad is immersible but, when placed in a bathtub, may begin to sing.

Section B: Service

• If your dad begins to show signs of wear, don't tell him. He might join the Hair Club for Men or get pectoral implants.

• If your dad is injured, seek treatment immediately, or he will attempt to treat himself by "walking it off." Dads who have removed their legs with chain saws have been seen hopping down the street, trying to walk off the injury.

• If your dad gets a cold, place him in bed and have the family hover around him with looks of concern. Dads like that.

Section C: Uses for Your Dad

• Your dad is versatile. He can function as a teacher, nurturer, entertainer, protector, adviser, confidant and cash machine. He's similar to your mother but with more of a propensity to burp.

Dads are particularly useful for:

• Killing spiders in the home.
• Reaching stuff on the top shelf.
• Teaching you how to spit.
• Throwing you long distances in the swimming pool.
• Creating small, recreational explosions.
• Allowing you to go an entire weekend without washing your face.
• Building elaborate Lego structures and then thinking of cool ways to knock them down.

Note: Certain dad models also can be relied upon for fashion advice. To determine whether you have a fashion-competent dad, check his closet. If you see neckties dating from the Reagan administration, seek advice elsewhere.

Section D: Troubleshooting

If your dad begins a sentence with "Don't make me stop this car" or "As long as you live under my roof," you're in trouble.

Section E: Accessories

Many accessories are available for your dad. These include but are not limited to:

- large-screen televisions
- golf clubs
- power tools
- fancy electronic devices
- season tickets to something
- a Dodge Viper

Providing accessories for your dad is particularly appropriate on Father's Day, but there is no need to limit yourself to that occasion. The more the better.

—Sunday, June, 18, 2000

Mom's lessons: reading, religion and relationships

I'M 45, AND MY MOTHER STILL TELLS ME not to go outside with wet hair.

This doesn't bother me a bit. I know a lot of people who no longer have mothers, so I count myself lucky to have one who is present and still functioning in full mom mode.

If I were her, though, I'd relax a little.

Once kids reach their 40s, they need less supervision. If we haven't done anything to bring shame on the family by now, the chances of our getting thrown into prison or mixed up in politics are slim. I think Mom did well.

In fact, I'm imitating her now as I raise my own kids. Not in every little detail — I don't try to sell the kids on Lawrence Welk, for heaven's sake — but in matters of significance.

Every mother gives her kids gifts, and not the kind purchased in a store. The gifts I'm talking about are the influences, interests and environment that shape a child's life. I've been contemplating this as Mother's Day approaches.

Everyone's gift list will be different. Here's what my mother has given me:

• **A love of reading**. My mother tells me people used to laugh at her because she'd sit with me on her lap when I was an infant and read to me. She did it for pleasure; this was before experts came along to announce that reading to babies is good for their brains. Nevertheless, it was quite a gift.

As I grew, I noticed that my mother and father read things, mainly the newspaper. So I started doing it, too. In fact, I read so much, people thought I was a little weird. Which I am, but that's OK — I've found a job where I can put it to good use.

I watched too much television as a kid, but without this love of reading, I would have watched even more. Generally, I think the more you read, the smarter you get, and the more television you watch, the dumber you get. So reading held me at equilibrium.

If I hadn't read so much, I would not have developed a love of words and become a writer. The world would have suffered. Not because it would have been deprived of my prose, but because I would have been forced into something where I could cause real harm, such as auto mechanics or bridge building. People who've seen me attempt to build or fix something are always grateful that I'm a writer.

• **A stable household**. My mother and father stayed married until he died. They wanted to, and they were living in a time when changing spouses wasn't the norm. Still, what a gift that stability seems today.

I was a bit of a worrier as a kid. I worried about the Cuban Missile Crisis at age 8. (There are some drawbacks to kids reading the newspaper.) I'd hate to think how I would have handled the anxiety overload of a dissolving family.

My mother didn't know that she was acting as a marriage counselor when she stayed married for 32 years. But where else do you learn what "for richer, for poorer, till death do us part" really means if not in your parents' house? I learned it in mine, my wife learned it in hers, and we have a marriage going on 24 years old.

• **A spiritual life**. My mother made us go to church, and skipping it wasn't a possibility unless we were ill. And to prove illness, we had to have a fever. There's no telling how many viruses we spread through the congregation in our formative years.

The world owes a debt to every mother of every baby-boomer kid who was dragged to church this way. As self-absorbed, whiny and materialistic as baby boomers are, can you imagine how much worse we would be if some of us hadn't been hauled into pews and made to contemplate things larger than ourselves? Talk about a gift.

I'm giving my kids the same gift, although at this stage they'd be just as happy to refuse it. This doesn't worry me too much. I operate on the germ theory of religion — I figure if you expose the kids to it long enough, it'll catch.

This is not an exhaustive list of Mom's gifts. I haven't even mentioned honesty, music and a great chicken recipe. But I don't want to praise her too much and risk making her complacent. She still has grandchildren to watch out for.

—Thursday, May 6, 1999

Time flies when you're having fun watching kids grow

MY SON GRADUATED FROM HIGH SCHOOL on Sunday, about two days after he was born.

I had plenty of notice that he was growing up, sometimes in ways that left me literally breathless.

One day during the summer before he entered ninth grade, we went running.

He took off at a fast pace, and I chuckled at his impetuousness. Oh, the foolishness of youth, I thought: We'll see how fast he's running at the end of 2 miles.

Well, at the end of 2 miles he was still running pretty fast. At least he seemed to be from a quarter-mile behind.

OK, wiseguy, I thought: You won, but it was my DNA that made your victory possible.

Don't I sound just like other parents? We cling to whatever thread still connects us to a growing offspring who seems to be moving away at frightening speed.

The distancing starts immediately, by the way.

I remember one thought from the moment my son was born: Wow, he looks so — separate.

Babies burst on the scene too fast, if you ask me. After nine months during which they remain somewhat abstract (at least to a dad), they suddenly emerge fully formed and ready for action.

I found him just a little startling.

In my egotism, I had imagined more of a miniature version of me — perhaps even with a little mustache and glasses. (My mind's eye insisted, even for a girl).

The boy arrived instead as an independent operator. Other people said he looked like me, but I couldn't see the resemblance: He looked like himself.

He acted like himself, too.

He was much more sociable and self-confident than I was as a child.

Dropped off at preschool, he entered the room as if he owned it.

"Could you come back here and cling to my leg for a moment?" I'd say.

By elementary school, my son had become a bit of a mystery — as I suppose all kids do.

He was out of parental sight for long periods, requiring that his mother and I rely on reports from the field.

My advice to parents: Never miss conferences with teachers, who fill in large gaps of information.

A child's public behavior differs from his private behavior — if you're lucky.

I remember saying to a teacher, "You know, he can be a little headstrong and loud and messy..."

She finished the sentence: "...at home."

He feels safe at home, she said, so he tests his limits there, in a secure environment. He'll cut loose somewhere, she added; be glad he does it at home, where no one keeps a file on him.

I tried to recall her comments whenever his bedroom took on the appearance of postwar Europe.

When a child graduates from high school, his parents bask in the congratulations.

"Such a fine young man — you should be very proud," people say.

I am proud; I'm also hesitant to take much credit.

First, because parenthood never ends, I'm afraid that to accept too much praise would jinx the process, which is far from over.

Second, I'm not sure exactly why my son turned out as he did — wonderful (but I'm not bragging).

By early adolescence, if not sooner, most kids have figured out how little control their parents really have over them.

Then what?

Then the parents hold their breath, hoping that whatever internal guidance system has developed will carry their child through the risk-filled teen-age years.

And, suddenly, high-school graduation arrives.

Whew.

I feel dizzy, as if his entire childhood just rushed by me in a day and a half.

Whatever sadness I feel is mitigated by the presence of the 11-year-old.

We play basketball together, and before long she'll start beating me.

—Tuesday, June 4, 2002

Couple's hopes tumble during dryer-delivery crisis

WE KNEW THE RISKS.

If you buy a major appliance, you're going to have to endure the loss of freedom that comes with waiting for delivery.

Still, who would have expected a three-day hostage crisis?

Even during the worst episodes — the time the refrigerator arrived three hours late and with the door hinges on the wrong side, for example — we'd never faced confinement like that.

I suppose we shouldn't have been surprised. We have a history of difficult merchandise deliveries. I've seen babies arrive in the world with less fuss than it took to get a queen-size mattress to the house last year.

The problems always start with the delivery "window." You know — the merchandise will arrive sometime between 8 a.m. and 1 p.m. Friday, they tell you. Can't be more specific, they say. Impossible.

In a society in which people work in the daytime, being told that something will arrive at your house between 8 a.m. and 1 p.m. Friday is only slightly more useful than being told it will arrive sometime between January and March 2004.

In the 21st century, shouldn't we be a little more precise? We live in a country that can hit an ox cart with a missile from 500 miles away, but we can't pinpoint when a washing machine will arrive in Reynoldsburg.

This much the store will tell you: The truck leaves the warehouse at 7 a.m. Helpful information? No. In all my years of buying appliances, I've never known a store whose truck didn't leave the warehouse at 7 a.m.

The information I covet is when it will get to my house. But this is unobtainable.

Apparently, when a delivery truck leaves the warehouse, it passes into a netherworld, beyond contact with humanity. People atop Mount Everest call their cousins in New Jersey just to say hello, but a delivery truck can't communicate with customers across town. Once it leaves the warehouse, a delivery truck is like Apollo 13 on the dark side of the moon.

And so our captivity began.

Day 1 dawned bright with promise. We had a morning delivery window. We were confident that we could make it all the way to noon if necessary. In our optimism, we even dared to imagine the dryer arriving at 10 a.m.

The dryer arrived at 2 p.m.

The store, knowing we had done hard time, waived the delivery charge. This would have been more comforting had the dryer that arrived late actually been the one we wanted.

When you wait all day and get the wrong dryer, something happens to your spirit.

I'm not particularly excited by dryers, but a sense of anticipation does build. You're a little keyed up. You get an urge to wash a load of socks, just so you'll have wet stuff ready when the machine arrives.

There was no hope of receiving a substitute dryer on Day 1, we were told. Why, the warehouse must be 10 to 15 miles away. Journey there on this primitive, multilane interstate highway system we have in central Ohio? Impossible.

Day 2 began darkly. We couldn't get a window. Why? Procedures, protocols, planetary alignment — I don't know. But they swore they would load our dryer toward the back of the truck in preparation for unloading it early. And, as we know, the truck leaves the warehouse at 7 a.m.

The dryer arrived at 1 p.m. It had a big dent.

Look, I know a dryer isn't a Lamborghini. People don't swoon when you show them your sleek, shiny laundry appliance with extended cool-down cycle and moisture monitor. But it was the principle of the thing. We'd done our time, and we deserved a pristine dryer.

Day 3 was a Sunday. No deliveries. We existed in a strange, in-between place, free but not free, like inmates on furlough.

On Day 4, the truck returned, within the promised window. Our eyes unaccustomed to the glare of a sun we'd seen only rarely in the past few days, we emerged from the house half-afraid to believe that our captivity might be over. But there it was, the right dryer, undamaged and ready for installation.

By now the store had knocked more than $100 off the price to atone for our time in captivity. It occurred to us that three more botched deliveries and we could get the thing for nothing.

But, no, we accepted the dryer. You can't put a price tag on freedom.

—Saturday, November 17, 2001

Two

My Town

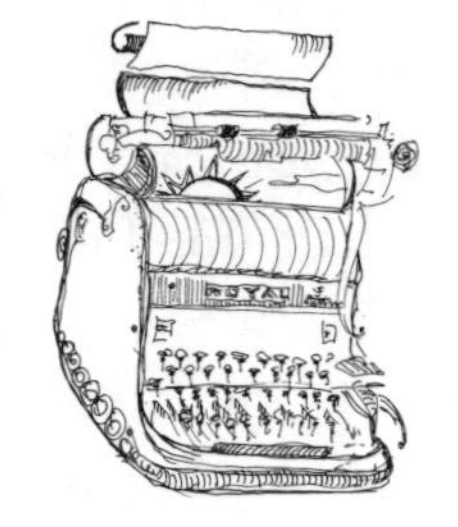

Jerry stays there, the rest move here and yada, yada, yada

To: Seinfeld cast

From: The producers

Re: Here's the script for the final episode. Working title: "Columbus Daze." It is ABSOLUTELY CONFIDENTIAL.

(The episode opens in Jerry's apartment. George lies on the couch, in shock. Jerry stands nearby.)

Jerry: George, you can't leave New York. You're self-centered and amoral with delusions of grandeur. You blend into a crowd here.

George: Mr. Steinbrenner doesn't care, Jerry. He said the Yankees' triple-A farm club in Columbus, Ohio, needs me. He's sending me to a place where I'll be a front-office big shot. He's sending me off to fulfill my potential.

Jerry: He's sending you down to the minors, George.

George: I know, I know.

Jerry: Columbus, Ohio. George, you're doomed. You don't have the casserole skills for Midwest living. Have you ever even been to a potluck dinner?

George: Never.

Jerry: What about coping with nature? Do you know how to apply broadleaf weed control?

George: No. I've never even mowed a lawn. I hate nature, Jerry! That's why I live in New York.

(Elaine enters, tearful.)

Jerry: What's with you?

Elaine: Mr. Peterman called me today. (Sniffling) He's selling his catalog to the people who publish the Victoria's Secret catalog. He recommended me for a prestigious job with those people. (Sobbing) They're impressed with my ability. (Wailing) They want to hire me at a salary of $300,000 a year.

Jerry: Financial security and a renewed sense of self-worth. You poor thing. Who wouldn't be devastated?

Elaine: You don't understand. In order to take the job, I have to move to a location so wild and remote that everyone drives sport-utility vehicles. I have to move to.... Columbus, Ohio.

George (leaping off the couch in excitement): Columbus, Ohio!

Jerry: Columbus, Ohio? Victoria's Secret is in Columbus, Ohio? I didn't even know they had lingerie out there. I thought everyone slept in bib overalls.

Elaine: No, that's Indiana.

George (grabbing Elaine by the shoulders): Elaine, listen to me. I'm moving to Columbus, Ohio, too!

Elaine: You are?

George: Mr. Steinbrenner is sending me there to work for the Columbus Clippers!

Elaine: What's that? A barber school?

George: It's a baseball team. And I'll tell you something else, Elaine. I've been reading about Columbus, Ohio. It's not like you think. It has a symphony. It has art. It has opera. It has dime-a-dog night.

Jerry: Now the attraction becomes clear.

(Kramer bursts into the room, wearing dark glasses and a wig.)

Jerry: Kramer! What's going on?

Kramer: I'm entering the federal witness-protection program. This is goodbye.

Elaine: Kramer, are you in some kind of trouble?

Kramer: I ratted on a mafia kingpin.

Jerry: Oh, no. Did you accidentally witness a gangland slaying?

Kramer: No. I was in the express lane behind a don who had more than 10 items in his cart. I complained, he started a ruckus, the cops came, and the next thing I knew I was singing like a canary to the FBI. They said they'd have to change my identity and relocate me for my own safety.

George: Where are you going?

Kramer: Columbus, Ohio. I'm going to become Brutus Buckeye, so I can wear a disguise at all times.

Elaine: Who's Brutus Buckeye?

Kramer: He's an ungainly character whose bizarre antics some find amusing.

Jerry: I thought you said you were changing your identity.

(Elaine, Kramer and George all suddenly turn and look sadly at Jerry.)

George: Jerry, come with us. We can be neighbors in Columbus, Ohio. We'll all vote Republican, have cookouts, use mulch.

Elaine: What's mulch?

Jerry: No can do. Comedians have to live in New York, where the grim realities of life feed our creativity. But I'll visit you often in Columbus, Ohio. Assuming it has an airport.

George: Enough with the Columbus jokes, Jerry. You're talking about my home.

—Thursday, April 23, 1998

Stadium cakes, cholesterol cows and the world's most wondrous hucksters

THE OHIO STATE FAIR SMELLS. It gets all gussied up for its visitors but still looks seedy. It is loud and a bit hucksterish. It is bad for my cholesterol level. And I love it.

I'm from a grimy little industrial city in Pennsylvania, so my affection for the fair has nothing to do with rural roots or state loyalty. I just think it's the greatest entertainment spectacle in Ohio. If admitting that disqualifies me from being an urban sophisticate, so be it.

The people who run the fairgrounds insist on the name Ohio Expositions Center, which I think indicates they don't realize what it is about the place that makes it charming.

Plunked right down in the middle of a big city striving to be polished and refined is a place that couldn't be if it tried. Not with that enormous walk-through OHIO at one entrance and a 15-foot fiberglass cardinal at the other. You park in a dusty field and ride a tractor-pulled wagon to the gate. Inside, the streets are named after live politicians, and you have to watch where you step. One of the attractions is a cow sculpted in saturated fat. As far as I'm concerned, that's all just fine. We have an abundance of sterile, contrived shopping malls, subdivisions and theme parks; there's only one gritty old fairgrounds.

That's why I'm a little worried about the motels and amphitheaters and fancy names attaching themselves to the fairgrounds. I'm afraid the fair organizers will go upscale on us one of these days, putting sleek facades on all the buildings and piping music into the restrooms. I'm afraid the fairgrounds are going to get a concept and lose the endearing appearance of having been made up as we went along. What a shame that would be. I like the place as it is — a jumble of mismatched buildings spread over too much land. There are surprises lurking in those buildings, and the fair doesn't spoon-feed them to you. You have to walk a lot and keep your eyes open.

The fair is an 18-day assault on the five senses. When I think about the fair, two images always come to mind: hundreds of lacquered dead cockroaches and a cake made to resemble a bucket of fried chicken. I saw the cockroaches one year in the Cox Fine Arts Building, where the more avant-garde artists display their work. The cake I saw in the DiSalle Arts and Crafts Building, where there is always fierce competition among bakers to see who can make the most bizarre desserts. (The Ohio Stadium cake, complete with C deck, is a recurring theme there).

The fair, you see, is a place where the pent-up talents of everyday people finally get released. This August opportunity to show off sends a flood of glory-seeking Ohioans rolling into Columbus. They come from all over the state with their home-canned pickles, their needlepoint portraits of James A. Rhodes, their prize squash and their carefully bred animals. Oh, those animals — chickens with punk hairdos, exotic rabbits with ears like beagles, hogs the size of a subcompact. People in this state do some fascinating things with genetics in their spare time.

I never go see the big-name entertainment at the fair — who has time? I want to see real people doing things I don't understand, like teen-agers in rhinestone outfits making their horses prance just so, or kids with sticks prodding their balky sheep around a show ring. I want to walk out of a building and smack into a cadre of elementary school baton twirlers, uncoordinated but so dead-earnest I just hope they don't hurt themselves throwing those metal rods around.

The state fair entertains me without trying.

One of the best shows going is put on by the people who hustle cookware. The fair is shamelessly commercial, of course, but these pot- and-panhandlers at least make you smile while trying to part you and your money. They never prepare anything that takes less than a dozen ingredients, so their spiels always involve a flurry of foodstuffs expertly assembled into an eye-catching meal in 60 seconds, with the help of whatever miracle product is being pushed. They've never persuaded me to buy anything, but I always leave thinking I should toss a coin into their chefs' hats.

We don't need to improve the fair with bigger stars, brighter lights or better buildings. I say let the people express themselves. Give us more crazy cakes, more insect bodies artfully displayed, more tiny trains clacking through mock towns, more salesmen trying to make a living. Give us a whole herd of butter cows.

No, I don't sneer at the fair. Sometimes I sneer at its promotion, which seems bent on convincing me it's Las Vegas on 17th Avenue, but I think the fair itself is amazing. It's like those doughy elephant ears they sell on the midway — all the sugar coating in the world wouldn't help if the ingredients didn't all come together and rise of their own accord into something people want.

—Sunday, July 31, 1988

Game-day 'PMS': Pre-Michigan Syndrome stresses out OSU fan

EXCUSE ME IF I SEEM ON EDGE TODAY, but I've got PMS. And I suspect I'm not alone.

Pre-Michigan Syndrome hits every year as The Game approaches and the pressure builds to hate Michigan with the kind of fury that comes naturally only to rabid Buckeye fans.

I will describe my condition in the hopes that it might help other sufferers.

First, there's the constant feeling of inadequacy. Being expected to hate an entire state is a terrible burden.

I wake up in the middle of the night, worrying. Have I hated enough? Did I remember to despise Muskegon? What about that glimmer of affection I felt for Sault Ste. Marie last week?

Then, there's the guilt. You should have seen how I agonized over the possibility that I cost Ohio State the game in 1996.

I arose before dawn the morning of The Game last year because I had neglected until then to abhor the Upper Peninsula. I went off to the spare bedroom, closed the door and started loathing everything between Manistique and Peshtigo.

Well, it just goes to show why attention to detail is so important in football. Peshtigo is in Wisconsin. All that hate, wasted.

I walked around for weeks after The Game, fearing that I was responsible not only for the Buckeyes' loss but also for every traffic accident that occurred in northeastern Wisconsin that day.

Lately, I've been retreating into fantasy to cope. I imagine that I've finally become adept at hating Michigan by inventing one of those virtual pets the kids are so crazy about, only it has the state of Michigan instead of a little dog on its screen.

Every half hour, the MichiGiga Pet beeps, signaling that it's time to commit another act of cruelty. I punch a button and raise the little Michigan's taxes, double the size of its state legislature or elect Marv Albert governor. Finally, I've focused my rage.

But then the depressing reality returns. Even if I succeeded in hating the state, I would still be struggling to hate the University of Michigan's colors.

Hating colors isn't easy for me. I can work up some passion against blue, though if the truth be known it's really pastel blue, not Michigan's much richer hue, that I'm not fond of.

But the maize is simply impossible to hate. It reminds me of corn on the cob, and who on Earth can hate corn on the cob? The best I can manage is mild annoyance for the kernels that get stuck between my teeth.

I thought I might think worse of maize if I arranged corn kernels into a portrait of Saddam Hussein (I used black beans for the mustache) and meditated on it.

Try as I might, however, I couldn't produce a convincing corn portrait of Saddam. It kept coming out looking like Captain Kangaroo, so I concentrated on hating him instead. If the Captain commits a key turnover today, you'll have me to thank.

PMS also leaves me exhausted and weak from wrestling with a vexing moral dilemma.

All citizens of Columbus are expected to offer, on demand, a prediction on the outcome of The Game. Unfortunately, I am terrible at these predictions. If I say OSU 21–17, it's almost certain to be Michigan 48–0.

So as a public service, I really should predict a Michigan victory. But this would be socially incorrect. I'm already under enough pressure for failing to have a little OSU flag flying from a holder clamped on my car window. (I don't like the funeral-procession look it gives my car, OK?)

In 19 years in Columbus, this is by far the worst case of PMS I have had. It's the bowl picture that aggravates it this year.

Hating Michigan alone won't do this season. I also have to think ill of Nebraska, Florida, Tennessee and I can't remember what other states in an effort to get the Buckeyes an invitation to something sexier than the Carquest Bowl.

I've never seen such a complicated scenario. It occurred to me last night that if OSU and Michigan tie, Florida State loses, Penn State narrowly escapes defeat and Nebraska forfeits its entire schedule, the Orange Bowl will be left with no choice but to pit the first runner-up in the Miss America contest against Vice President Al Gore for the national championship.

The only good thing I can say about my PMS is that it's almost over. I sit here on the morning of The Game knowing that, no matter who wins in a few hours, I will find relief.

I don't, thank goodness, suffer from Post-Michigan Game Distress Syndrome. That afflicts only the really loyal fans, and only when the Buckeyes lose. I'd like to predict that they will, because, of course, that means they won't. But I think I better just go lie down.

—Saturday, November 22, 1997

Leatherlips' curse has nothing at all to do with rain

IF IT SO MUCH AS DRIZZLES FOR 15 MINUTES this week, people are going to say it's the curse of Chief Leatherlips come to haunt the Memorial Tournament at Muirfield again.

As the story goes, Leatherlips, the old Wyandot chief, is angry with Jack Nicklaus for building a golf course too close to a Wyandot burial ground. So he drenches the tournament every year.

I think Leatherlips was more clever than that.

The way I figure it, Leatherlips and the deceased chiefs of many other tribes, already possessing a long list of grievances and disgusted by the way the land was being used, assembled decades before Nicklaus was even born to hatch a curse. It would bedevil the descendants of white settlers for generations. And it wasn't rain.

Here's how it I think it went:

Leatherlips gathered his colleagues and told them he wanted to curse the white settlers with a game.

"A game?" they said. "What kind of a curse is that?"

Imagine this, Leatherlips said. On one hill rests a ball no larger than a bird's egg. On another hill far away is a hole no wider than a mole's tunnel. Those playing the game must use a stick to hit the ball into the hole in only three or four swings.

The other chiefs hooted. "Nobody is going to play a game like that," they said.

People will become obsessed with it, Leatherlips assured them. They'll play hour after hour, journeying up and down hills, through thickets and into creeks in pursuit of the ball.

"Won't they get lost or eaten by wild animals?" the chiefs asked.

Their clothes will protect them, Leatherlips explained. Their lime-green pants, white shoes and turquoise visors will make them easy to spot from a distance but unappetizing to predators.

"Won't they eventually get discouraged and quit such a difficult game?" the chiefs asked.

No, because every once in a while they will play well and imagine that they are finally starting to master it, Leatherlips said. But then, the next time they play, they will find that all their skills have suddenly left them. Many balls will be lost in water, and they will hurl their sticks in sorrow and despair.

"Will everyone play this game?" the chiefs asked.

No, Leatherlips said. Some will wait at home for the players to return and provide lengthy, detailed, partially truthful accounts of the day's contest.

"Why should those at home suffer so much?" the chiefs asked.

That's how it must be, Leatherlips said. Then he described more strange behavior. To play this game, the participants will have to maintain broad expanses of finely cut grass as far as the eye can see.

"Grass?" the chiefs chortled. "In Ohio? In the spring it will grow wildly out of control and choked with weeds. In the summer, the sun will wither it into a shriveled brown mat. They would have to spend hours a day obsessively tending it in a grandiose attempt to bend nature to their wills."

Now you're getting the picture, Leatherlips said.

The other chiefs began to warm up to his idea. Soon they were contributing suggestions of their own and laughing uproariously.

Make them use words like birdie and bogey in serious conversation, one chief said.

Have them traverse the hills in comical little carts that make their bellies jiggle, another suggested.

Convince them that they can play only in absolute quiet, so that a single chirp from a robin can ruin the entire game, said a third.

Leatherlips began to regret introducing his idea to such a large group. Curses designed by committee quickly can become bizarre. But he managed to accommodate all their suggestions and presented the finished curse to his colleagues one day late in May.

Leatherlips said soon he would put the curse into effect and they all would watch in pleasure as the victims struggled to play the game.

"We'll have to wait a few weeks, of course," the other chiefs said.

Yes, Leatherlips said, looking up at the overcast skies. No one would attempt to play the game at this time of year.

—Tuesday, May 26, 1998

Training camp gets fans ready for beer-can curls

OHIO STATE'S FOOTBALL PLAYERS HAVE BEEN doing wind sprints and hitting the blocking sled in preparation for the season that begins Sunday. But how do fans get ready?

Here's a report from Buckeye Fan Training Camp:

Sit tight — Two injuries were reported yesterday after coaches ran the dreaded Ohio Stadium seat drill, in which 24 fans practice squeezing themselves onto a bench 18 inches long.

Dutch McGillicudy of Westerville will be out at least two home games with cracked ribs. Slim Simpkins of Whitehall briefly lost consciousness when he was sandwiched between Big Mama Smith of Newark and Monument Monahan of the South Side.

"You hate to see anyone get hurt, but that's how they're going to have to sit when the season starts," said Coach Aurelius "Homer" Simpson.

On the go — Assistant coach Flo Fabuloso said she was impressed with how the rookie squad handled restroom sprints yesterday.

Tina Gazina of Toledo clocked the best time with a 48.3-second dash from C deck to the ground-level women's restroom near Gate 18.

"But then coach made me wait in line for 11 minutes when I got there," she said. "It was tough because I'd just had two beers. But I have to take my bladder control to the next level."

Feeling their pain — Grief counselors will be on standby at sports bars after the Penn State and Michigan games this year, Simpson confirmed.

"We hope we don't need them, but it's better to be prepared. These are Buckeye fans."

Slow learners — The university announced that 48 more fans have been declared academically ineligible. Officials refused to elaborate, but a source said at least 20 were dropped because they kept shouting the wrong letter during the O-H-I-O cheer.

"Another guy refused to do the wave because he said he couldn't swim," the source said. "You have to wonder how these people got out of high school."

Hot dog — The talk around camp this week has been about the performance of a new brand of bratwurst.

"It's stepped up big time," said tailgate testing coordinator Sear O'Malley.

The new brand is Big Log O' Hog out of Parma. It's expected to see extensive grilling time during the home opener against UCLA, O'Malley said.

Meanwhile, a Yellow Springs product, Self-Righteous Farm's Holistic Vegetarian Brats With Calcium, has been a disappointment.

Get a grip — The coaching staff has put special emphasis on avoiding fumbles this year.

"It's no secret that last year we had trouble hanging on to the nachos," Simpson said. "That can't happen again. If you can't make it back to your seat without dropping congealed cheese sauce into people's hair, you don't belong in Division 1–A football."

Yesterday, the coaches made each fan run the tire drill while holding a 24-ounce

Coke, a large box of popcorn and a hot dog with onions and sauerkraut.

Injury update — Charlie Charcharian of Chardon caught his pants in a turnstile during entry practice. Coaches were relieved to learn that Charcharian was not injured, although his left hip pocket needed seven stitches.

Moose Wallaby of Delaware strained a vocal cord on the "land of the free" part of *The Star-Spangled Banner*. Coaches said he may have to drop down an octave for the first home game.

Jimmy "The Vent" Jimmerson of Reynoldsburg bit off his own finger while practicing protests of questionable officiating. Simpson said he appreciated the effort but discouraged other fans from following Jimmerson's example. "It's hard to cheer and hemorrhage at the same time."

Briefly — Pregnant fans are being urged to go to the hospital rather than giving birth in stadium restrooms this year. "If it's a close game, we might let them stay, but generally we'd prefer they find a maternity ward," an OSU official said.... The Scarlet Ranters, OSU's precision team of malcontents, looked sharp during hyperventilation drills yesterday.... Bob "West" Virginia of Obetz said he's been granted permission to bring his deceased uncle to all seven home games. "It took my uncle years to get season tickets, and he's not about to give them up."

—Tuesday, August 24, 1999

Hey, Hollywood! Columbus makes a perfect setting

I've been inspired by Mayor Michael B. Coleman's efforts to sell Columbus as a film location.

As you know, the mayor wants more movies and TV shows made here.

A *Dispatch* story on Monday discussed Coleman's efforts to woo Hollywood. In Los Angeles for the Democratic National Convention last month, he met with an NBC vice president, dined with Arnold Schwarzenegger and generally talked up Columbus as a dramatic location.

Had this been all he did, I wouldn't have been as moved. Arnold? Yeah, he stops by now and again and says he loves us. But when has he ever filmed a summer action picture here? If he really cared about Columbus, he'd bring a 40-member film crew and some explosives.

No, it wasn't the schmoozing that inspired me. It was the mayor's reference to *Providence*, the hit TV series set in the capital of Rhode Island.

"That show could have been named *Columbus*," the mayor said.

I love the approach, mayor. You want to get specific by asking Hollywood to imagine how existing TV shows and movies could be improved if they were set in Columbus.

Let me help. Here's how some well-known shows could be reworked:

CPD Blue — In the pilot episode, Sipowicz joins the Columbus police mounted patrol, endangering the vertebrae of the horses but striking fear in the hearts of those who would attempt an illegal left turn at Broad and High.

Sex and Grove City — The romantic adventures of sassy, sexy young women are set amid the glamour of strip shopping centers and fast-food restaurants in a Columbus suburb. In the first episode, Sarah Jessica Parker catches a man's eye by rear-ending his car in the parking lot at Wal-Mart.

Malcolm in the Midwest — Young Malcolm and his family find amusing ways to cope with the stresses of central Ohio life. In the first episode, Malcolm and his siblings get lost in a corn maze.

I think you'll agree that this is a promising start, but Hollywood has an insatiable need for new material. So I've banged out a few proposals for original scripts. The mayor is welcome to shop any of these around:

Knight Watchman — In this situation comedy, Bob Knight returns to central Ohio and takes over as basketball coach at a small high school, much to the chagrin of a hapless assistant principal assigned to monitor Knight's behavior at all times.

Bellisari — The TV medical drama gets a new look in this show about a young hero on whom the hopes of Columbus are pinned. Every week he impulsively dashes into risky situations, causing central Ohioans to flood emergency rooms for defibrillation.

Coup de Moss — In this summer action movie, Columbus is taken over by space aliens. Desperate to wrest control from the invaders, residents turn to a maverick school board member with a talent for driving the powerful crazy. First he sues the invaders and everyone else in the Alpha Centauri phone book. Then he lures them into a four-hour meeting at which he ostentatiously objects to everything they say. They decide to move to the suburbs.

The Sound and the Fury — Hilarity ensues when an outdoor amphitheater is built in the midst of quiet suburbs. The series focuses on the antics of the harried amphitheater staff, which must keep rock concerts under 50 decibels, sell overpriced bottled water, and make sure Phish fans don't urinate in the neighbors' flower beds.

Phone of Arc — In this religious drama, a young woman receives messages from heaven that inspire Columbus. But when the messages are disrupted, the woman can't get Ameritech to come fix the connection. She decides to become a utility regulator.

—Thursday, September 14, 2000

In quiet Columbus, first snow makes news crews scurry

YEAR IN AND YEAR OUT, THE TOP 5 television news stories in Columbus are:

5. It's raining.
4. It's hot.
3. It's cold.
2. Michigan beats Ohio State.
1. IT'S SNOWING!

Weather of any kind is a big story here, but snow thrusts newscasters into a frenzy that makes the doctors on *ER* look laconic by comparison.

The newscasters, of course, are merely reflecting the local understanding that, because we lack the hurricanes, earthquakes or mudslides that make other major-league cities interesting, we choose to be terrified by snow.

Still, I worry about inducing panic in newcomers. They could turn on the tube during a dusting of snow and conclude from the tone of the reports that the Earth is being pelted by comets.

To prepare newcomers for the winter onslaught of weather reports, I present this script of what snow sounds like on television:

Mr. Anchor: The big story tonight: Snow!

Ms. Anchor: Let's go live to Muffy Collagen, who is standing outside our studios for no apparent reason.

Muffy (in a bright yellow parka designed for Arctic expeditions): That's right, Hank and Sue, I'm standing outside in bone-chilling, 32-degree temperatures. I'm going to insert this ruler into the snow in a vivid demonstration of just how much snow has fallen. See? It must be at least three-sixteenths of an inch.

Ms. Anchor: Great TV visual, Muffy. The only thing that could make it better would be if the ruler were bleeding. Can you give us any obvious advice designed to make viewers think we care about them?

Muffy: Hank and Sue, police tell me that drivers should exercise caution. They say that when temperatures fall to freezing, moisture in the air crystallizes into snow, which often collects on roadways and makes them slippery. So, please, be careful, because we wouldn't want to have to send camera crews rushing to the scene of gruesome accidents with gripping scenes of passengers being cut from the wreckage of their flaming vehicles.

Mr. Anchor: Good advice. Thanks, Muffy.

Ms. Anchor: Yeah, thanks, Muffy. Now let's go live to Rick Follicle at the city's snowplow garage. Rick, are the city's cleverly named "snow warriors" ready for this storm?

Rick: Hank and Sue, the snow warriors started salting freeway overpasses on Aug. 15 just to make sure they got a jump on this storm. In fact, the Spring-Sandusky

Interchange has so much salt on it, people are calling it the Hot Pretzel.

Ms. Anchor: Rick, can you introduce an element of suspense into your report that will carry us through the upcoming commercial break?

Rick: Hank and Sue, officials tell me this giant mountain of salt behind me represents the city's entire supply. Once it's gone, Columbus could descend into a frozen hell.

Ms. Anchor: When we come back: How long will the salt last?

(Commercial break)

Mr. Anchor: For more of our Special Team Report on the big storm, let's go to Linda Getreal at City Hall.

Linda: Hank and Sue, I'm here with Mayor Coleman. Mayor, does the city have enough salt?

Coleman: Yes.

Linda: Back to you, Hank and Sue.

Ms. Anchor: Thanks, Linda. And now let's go live to our Lifesaver Weather Central High Command, where our certified meteorologist Storm Varning has our colorful maps with twitchy images of clouds. Storm, can you stand in front of a map and make sweeping gestures while uttering incomprehensible jargon?

Storm: Yes, Hank and Sue. We've got an area of high isometric pressure here, as indicated by the counterclockwise movement of thermocouplers over there. Cumulative cloud formations, over there, are giving us a wind-chime factor of 27 degrees Canadian, here. Meanwhile, upper-air disturbances are forcing the jet stream to circle LaGuardia, which I won't point to because I just strained my back imitating cloud rotation. Back with my forecast after this.

Mr. Anchor: When we come back: Find out how bad it'll be.

(Commercial break)

Ms. Anchor: Well, the snow continues to fall. Let's go back to Storm Varning for his exclusive Lifesaver Weather Forecast.

Storm: Cold tonight with flurries.

—Saturday, December 2, 2000

Columbus may be big, but it still feels small

I KNEW THIS WOULD HAPPEN. The census figures come out, Columbus proves to be the 15th largest city in the nation, and instantly we begin whining: We don't feel as big as we are. We don't have an image. We're misunderstood. We're overlooked. Nobody loves us.

Get a grip, people.

There are many good reasons why Columbus, a big city, doesn't seem like a big city. This will not change in our lifetimes.

Here are the reasons:

• **Weather**

Traditionally, major-league cities have had major-league weather. Consider Los Angeles weather: earthquakes, mudslides, brushfires, Pacific waves majestically pounding trophy houses into rubble.

I'm sorry, but our weather just doesn't compare. Miami has Hurricane Andrew. San Francisco has the Loma Prieta earthquake. We have overcast skies with a chance of drizzle. We lack drama.

Don't try to blame this on the media, either. The local TV stations are relentless in their efforts to dramatize our weather. Channel 10 will break into regular programming to report a puddle forming in Hilliard. Channel 4 actually makes snow. (Channel 6, with a smaller budget, just scatters Styrofoam peanuts.)

What are the prospects of our weather acquiring the menace that lends such an allure to L.A.? Dim. We may be able to create the illusion of weather with some simple steps, such as planting trees on the slant. Otherwise, I see little hope, unless global warming gives us a coastline.

• **Sin**

We have sin, but not the kind that attracts tourists. Being reserved Midwesterners, we prefer to sin in our homes, which are all on cul-de-sacs, where out-of-town visitors are unlikely to roam in search of excitement.

Personally, I prefer home-based sin, but it's incompatible with the image of a big city. Fly into Las Vegas sometime. You can commit three or four sins just walking through the airport. Here? The closest you'll get to sin is cursing at the complicated escalator arrangement at Port Columbus.

Without marketable sin, Columbus lacks that whiff of danger that makes you want to put on a bulletproof vest and stroll the streets of Miami.

Could we persuade people to change to a more recreational sin pattern? Perhaps in a few generations, if we start teaching it in the schools. But we face enough of a struggle just getting everyone to pass the fourth-grade proficiency test.

• **Celebrities**

We like to brag about famous people who are from Columbus, which is nice but irrelevant. Plenty of famous people were born in Columbus. The problem is that they moved.

Do you see the chicken-and-egg issue here? Famous people will live only in cities that seem big, but cities can't seem big until famous people live there.

Worse, the famous people that we do have tend to be subordinate to other cities' famous people. Hence, Jack Hanna appears on David Letterman's show, not vice versa. This only reinforces stereotypes.

We could form a committee to attract more famous people to Columbus, maybe by offering tax abatements. But I'm not sure it would be in the public interest. Once the novelty of famous people wears off, you realize all they do is tie up traffic.

• **Cultural references**

Big cities are cultural touchstones. Their names turn up in food (Philadelphia cheese steak), songs (*New York, New York*), movies (*Sleepless in Seattle*).

One of our rare moments in the cultural-reference limelight was when the book *Goodbye, Columbus* came out. Alas, it was set in New Jersey.

Can anything be done? Yes, we could change the name of the city. Almost nothing rhymes with Columbus, so we rarely find ourselves used in song lyrics. I doubt, however, that city leaders would want to abandon the name, as it would render several statues irrelevant.

Our best hope may lie in the area of diet, where we have gained something of a reputation for being well-fed. We could lobby to get the term *Columbus-sized portion* into the lexicon. Then again, I'm not sure that's the meaning of big city we're aiming for.

—Tuesday, April 10, 2001

Party on, Buckeyes; you've earned it

OHIO STATE UNIVERSITY WAS NAMED THE eighth-best party school in the nation in the annual survey by *Princeton Review*. Ohio University, meanwhile, fell out of the Top 20. Here's an almost-factual report on campus reaction:

COLUMBUS, Ohio — Ohio State University suddenly finds itself among the elite party schools, but revelers at this Midwestern institution say they aren't letting the high ranking go to their heads.

"It's all just talk until the keggers start," said senior Bud Guinness at Gamma Ramma Lamma house.

"That's when we'll see who's willing to go out there every weekend and put their livers on the line. And you can't neglect the classroom either. Can you attend a 72-hour rave during finals week and still maintain a 1.6 GPA? We'll see."

OSU, No. 14 last year, climbed all the way to No. 8 in the *Princeton Review's* annual list of best party schools.

Tennessee was No. 1 in the preseason poll, followed by Louisiana State, California-Santa Cruz, Florida State, Colorado, Alabama, St. Bonaventure, OSU, Wisconsin and Florida.

"For us to be ranked ahead of a Florida school just shows how much we want it," said sophomore Mindy Hops-Barley from her hospital bed, where she was recovering after diving off a balcony into a trash can full of spiked punch.

"I mean, in Florida, partyers in January don't have to worry about being found the next morning frozen to a fire hydrant."

Eager to prove themselves, some Ohio State partyers have committed to two-a-day bashes when the party season resumes in early September.

"That's what it takes to party in Division 1–A," said one reveler who asked to remain anonymous because he couldn't remember his name.

This isn't the first time OSU has gone into a party season with high expectations.

Few here have forgotten the 1997 school year, when OSU jumped off to a quick start, only to run into midseason problems when a bad batch of happy hour hors d'oeuvres sent several hundred partyers to the toilets early.

"That was the night that 500 people were supposed to march naked down High Street at 1 a.m. while playing *Carmen Ohio* on kazoos and overturning every third car," junior Samuel Adams said. "When that fell through, we just lost party momentum. You could see it, man. The next weekend, there were people in the libraries."

The incident helped cement OSU's reputation as a party school that never quite "pours the big one."

Last year, however, partyers managed to throw several bashes that required police intervention, a fact that may have contributed to Ohio State's sudden jump to No. 8.

The emergence of OSU comes in the same year that its in-state party rival, Ohio University, fell out of the Top 20 for the first time in recent memory.

Veteran revelers at OU, with the reputation of a perennial party-school powerhouse, are still trying to figure out what happened.

The *Princeton Review* survey has some partyers speculating that a sudden fit of maturity had settled over the Athens campus.

"Maybe doubts about partying are starting to creep in," said OU senior Gerhard Grolsch. "I had some last winter, after I went to a party near campus and woke up the next day in Mexico, married and wearing someone else's clothes. It took me three days to prove I was an American citizen. I missed midterms and my grandmother's funeral. Was it worth it? I can't remember."

Still, some OU partyers scoffed at the idea of OSU eclipsing their school in revelry.

"It's crap," said OU junior Mickey Labatt while picking up trash as part of a work-release program for first-time offenders. "OSU the eighth-best party school? The eighth-best Tupperware party school maybe. But it'll never beat OU for hard-core celebration."

Administrators at both schools downplayed the party rankings, alleging that they are based on interviews done during wet T-shirt contests in campus-area bars.

—Saturday, August 25, 2001

Forebears slogged through many a gray winter

THE LONG, BRUTAL COLUMBUS winter approaches.

How will we bear up under day after day of 35-degree temperatures, drizzle and gray skies? How will we cope when snow flurries threaten? By taking inspiration from those who went before us.

I've unearthed excerpts from the diaries of central Ohio residents of long ago. When cold and darkness settle over the city, remember that these brave souls managed to survive the cruelly mild winters of central Ohio, and so will we:

• From the diary of Ebeneezer Polaris, early settler of southern Delaware County:

Nov. 28, 1823

Tenth straight day of overcast skies. Not that cold, really, just gray and drizzly. Oxen so depressed, they keep trying to migrate to the Southern Hemisphere.

Difficult to convince oxen that they are not migratory animals once they take a notion. Finally set the barn on fire, hoping the light will improve the oxen's mood or something.

Didn't work, and now I don't have a barn. Louisa seems angry with me: "Eleven kids to feed, and you're out burning down the barn to make the oxen happy?" Nag, nag, nag, I tell her.

We will survive.

• From the diary of Timothy Horton, one of the first doughnut merchants in Columbus:

Dec. 13, 1852

Huge storm swept in from the west last night, dumped at least an inch of snow. Only the very tips of the grass are visible under this suffocating blanket of white.

How will supply wagons get through? Running low on essentials — chocolate sprinkles, powdered sugar, cream filling.

I've been unable to make crullers for three days. People think I'm hoarding. Angry mob gathered outside the shop last night. Doc Jones said Columbus folks have high body-mass indexes, makes them testy during a cruller shortage.

"Pacify 'em with some jelly doughnuts and hope for the best," he said.

Forecast for temperatures in the low 40s tomorrow. Perhaps the mountain passes in Worthington Hills will clear enough for supplies to get through. We must persevere.

• From the diary of Jane Muirfield, Dublin housewife:

Jan. 3, 1874

They say there's going to be sleet today. Whole town in a panic. People are running from farm to farm, buying eggs, milk, whatever they can get. The quilting bee has been canceled, the church service is postponed, and Mrs. Wagner, the midwife, says women best not go into labor because she's not going out in this weather.

I can face anything as long as they don't call off school again. Being cooped up in the house with all 17 children will surely test my sanity. Last week, it got so bad, I sent the seven oldest ones outside to hunt buffalo.

"But, Mama, there are no buffalo around here," they said.

I told them I didn't care — they were to go outside and not come back until they had one. Little scamps finally got so desperate they glued some wool to a cow. But it kept them busy all afternoon.

These are difficult days, but we must carry on.

• From the diary of Alphonse Brutus, an Ohio State University professor whose gangly limbs, bizarre manner of dress and unusually large head are thought to have been the inspiration for Brutus Buckeye:

Dec. 18, 1899

The skies are leaden, and a damp chill has lain over the city for three solid days now.

How much longer can we survive in these gray conditions?

People are growing desperate to see color. The other day, I was walking on High Street when I noticed that a crowd had gathered. What were they gawking at? A boil on Phineas Ogilvie's nose. It had a reddish tinge that was quite arresting.

I fear for our transportation system. Snow flurries have begun, and almost immediately, wagon traffic is backing up on streets all over town. Drivers seem to take leave of their senses around here when snow falls. But I suppose it will get better when we begin using those new horseless carriages. Until then, we must remain resolute in the face of winter.

—Thursday, November 8, 2001

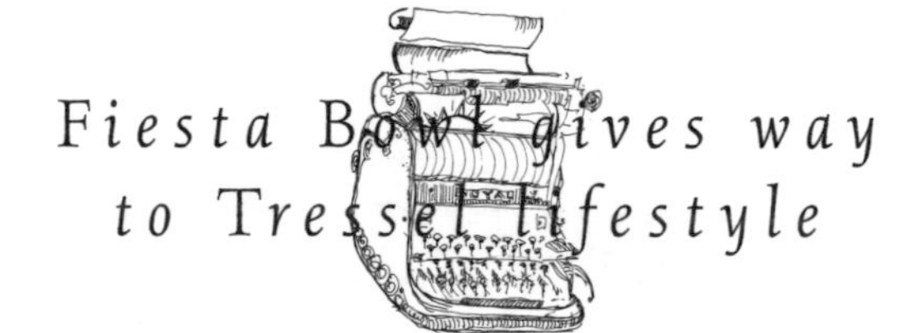

Fiesta Bowl gives way to Tressel lifestyle

A GLIMPSE INTO A SCARLET-AND-GRAY FUTURE: Tressel-mania washed over the nation like a storm surge after Ohio State won the Fiesta Bowl in 2003.

In the ensuing years, coach Jim Tressel influenced everything from fashion to politics.

A look back at how his style came to dominate American culture by 2008:

Fashion

The Tressel sideline look hit big in the fall of 2003.

The white shirt and sweater vest won predictable favor, but marketers were shocked by the popularity of the accessories.

Teen-agers donned oversize headsets and carried clipboards. Posing on corners, they pressed their fingers thoughtfully to their lips as if mulling a third-and-short situation.

When even Eminem dyed his hair salt-and-pepper to go with a George Hamilton tan, everyone knew: The hip-hop look had died. Tressel style ruled.

Language

Linguists noticed a change in sports talk after the Buckeyes won a second national championship in January 2004.

The sound-bite culture of brief, taunting, emphatic statements began to mix with a more nuanced form of expression.

Trash-talking gave way to "Tress-talking."

A graduate student studying the phenomenon miked opposing football players and recorded this exchange:

Trash-talking cornerback: "Your mama's a @#$%,&."

Tress-talking wide receiver: "My mama has worked very, very hard. I've seen effort; I've seen enthusiasm; I've seen dedication. Now, is she the best mama in the world? I'll let someone else make that determination. But I've been pleased with her progress, and I think she'll match up with your mama pretty well."

Merchandise

The Tressel action figure became the must-have toy of Christmas 2004 after the Ohio State winning streak reached 39 games.

"I swore the thing wouldn't sell," a toy-industry executive said. "I mean, what little boy is going to be satisfied with an action figure that reacts to danger by folding its arms and gazing analytically at the enemy?"

Social outreach

Tressel had become so influential by 2006 that people were asking him to solve their problems.

Though busy, the coach agreed to dispense advice on the Jim Tressel National Crisis Hotline.

A typical call:

Caller: "Jim, I'm a 103-year-old, blind quadruple amputee. I'd like to take up motocross as a hobby, but I'm afraid I'll fail."

Tressel: "My experience has always been that you just work and work and work to get a little better every day — not just in motocross but in all aspects of life as part of the Ohio State family."

Caller: "The what? I live in rural Wyoming."

Tressel: "We consider everyone on Earth, alive or dead, to be part of the Ohio State family."

Caller: "Bless you, Jim."

Politics

Tressel was elected president in 2008, carrying every state but Michigan.

Though popular with the public, he puzzled foreign leaders and frustrated the media.

He called North Korea "very physical."

He ended every news conference with reluctant reporters clustered around him, singing *The Star-Spangled Banner* to the TV audience.

He referred to America as "the best damn nation in the land."

New Yorkers smirked, and Californians giggled, but Ohioans beamed.

—Wednesday, January 8, 2003

Speaking the unspeakable: What to do when 'they' lose

IT'S A TOPIC NO ONE WANTS TO TALK ABOUT, but someone must: One of these days, the Buckeyes are going to lose.

Loss is inevitable in college football. It might come after a long period of deteriorating play or strike when least expected. It might come this year or next. But come it will.

The Ohio State fan's impulse is to avoid the subject, as if not talking about it will make it go away. But we must face the truth: Will Allen cannot intercept every last-second pass into the end zone with the game on the line.

Now is the time to make plans for helping Buckeye fans cope. After a defeat, they will be in a highly emotional state that can lead to poor judgment and bad decisions.

Here are tips on how to prepare:

• **Urge fans to discuss their health-care wishes.**

Do they want to be resuscitated? Don't wait until a fan is lying comatose in C Deck to attempt this conversation.

A good time to broach the topic is after a quadruple overtime victory. Such games give fans a glimpse of Buckeye mortality, although they may not admit it. Broach the topic gently, avoiding the word lose.

You might say, "If the Buckeyes should meet with misfortune one of these Saturdays, would you want heroic measures to revive you, or would you rather not be around to see them slip to No. 22 in the AP poll?"

• **Know the stages of grief.**

Fans will pass through distinct stages: shock, anger, rage, wrath, horror, hysteria, fury and fire-the-coach. They may call a talk-radio host and claim their kindergarten-age children could have devised better game plans.

Help them through this process by agreeing with everything they say.

• Don't say stupid things.

In their rush to console fans, friends sometimes turn to cliches that hurt more than help.

Don't say, "I know how you feel." Unless your sense of well-being hinges on the athletic performance of college kids in plastic hats, you can't possibly know.

Don't say, "It was part of God's plan." That's only true when the Buckeyes win.

Fans just want to know that you share their feelings. Simply say, "The officials screwed us," and leave it at that.

• Discuss disposal of property.

The worst time to make decisions about property is immediately after a loss. Typically, a distraught fan will, in the heat of the moment, put his foot through a 62-inch TV screen. Too often, it's not even his television.

Now is the time to discuss appropriate ways to express the devastation of a loss. Perhaps tailgaters could prepare a tub of sacrificial potato salad to be thrown into the Olentangy River.

It's especially important to have this conversation with high-spirited young fans, who have been known to overturn cars when they're in good moods.

• Don't rush them.

Grief is a process. Never tell fans to "just get over it." Some of them are still not over the 1976 Rose Bowl, for heaven's sake.

Expect pain to linger for weeks. Oh, fans may still tailgate, wolfing down bratwurst, deviled eggs, cabbage rolls, coleslaw, nachos and triple-fudge brownies as always.

But know that they carry an empty place in their hearts, if not their stomachs.

—Wednesday, September 24, 2003

Three Adventures

Glub, glub! He swims with the fishes, er, synchronized team

THE INVITATION FROM U.S. SYNCHRONIZED SWIMMING suggested sending a "macho man" from *The Dispatch* to try the sport. All the macho men were busy, so the editors sent me.

Organizers of the collegiate national championship, which starts today in Ohio State University's Peppe Aquatic Center, wanted to get a media person into the pool to demonstrate that synchronized swimming is a strenuous sport.

I really didn't need to be convinced that it's a sport. I'm 45. I think gardening is a sport.

Besides, the organization's Web site (in a single sentence) provided ample evidence of the athleticism required for synchronized swimming: "Swimmers might spend up to one minute under water without air."

In layman's terms, that's called a near-death experience, I believe.

Still, there are times when you have to put aside fear in the interest of making a fool out of yourself. I called Brian Eaton, a spokesman for the organizers, and told him I'd do it.

"Great," he said. "Maybe I'll put you in with the two Brazilians who are going to try out for the Olympics."

Great. Two preschoolers who excelled in Water Babies class would be closer to my level of ability. Yet Eaton said not to worry.

I worried anyway. There are two things in life that I do reluctantly: fly and swim. It's a genetic thing; my ancestors were land mammals.

Oh, I also don't dance well. So I was really in my element. Synchronized swimming is something like a Broadway production number performed in water. And sometimes people lift you into the air.

The roots of synchronized swimming go back to water ballets performed in the '30s and '40s. It's the only sport I know of that was inspired by Esther Williams movies.

I climbed into the pool about 1 p.m. yesterday, wearing nose plugs and weighted down by apprehension.

The Brazilians — Isabella and Carolina Moraes — were there, and I was relieved to hear them speaking English. I thought I might need to communicate important information, such as, "I need CPR immediately."

The other OSU team members who endured my thrashing about were Heather Moore, Mary Hofer, Katie Edwards, Tarin Forbes and Marietta Aruta. We were in water that was, oh, I'd say about 48 feet deep.

After entering the water, I willed myself to be calm. Otherwise, I might kick one of these athletes, knock her out of the competition and cost OSU a national championship. In this town, drowning would be preferable.

We started with a basic synchronized swimming move, the eggbeater. It involves rotating the legs in opposite directions to keep yourself up and stationary. I was down and drifting, so I don't think I quite got the hang of it.

Nevertheless, we pushed on. Most of the session went like this:

• One of the swimmers would deftly demonstrate a move, such as the ballet leg (extending the leg gracefully while floating on the back).

• I would attempt to imitate the move and quickly sink.

• After several tries I would finally succeed, but only because three or four swimmers were holding me up and manipulating my limbs into the proper position, as if I were Gumby (which I was, for all practical purposes).

The rest of the session went by in a blur, partly because I didn't have my glasses on and partly because I was exhausted.

Synchronized swimmers perform all their intricate routines while suspended in water. They don't touch the bottom of the pool. Neither did I. I hung on the side instead.

They were able to coax me to the center to attempt several more-advanced moves. These include the pop-up (launching yourself out of the water with both arms extended), the vertical (holding yourself upside down in the water with both legs above the surface) and the tower lift (being propelled out of the water on the shoulders of two teammates).

My pop-up was pathetic, my vertical was horizontal and my tower leaned. Other than that, I did well.

Meanwhile, the swimmers were gliding, spinning, inverting, extending and launching with impressive power and grace. It was like swimming with seven dolphins.

Mercifully, they let me out of the pool after about 20 minutes. I emerged spent. Behind me, I think I heard the sounds of synchronized laughing.

—Thursday, March 16, 2000

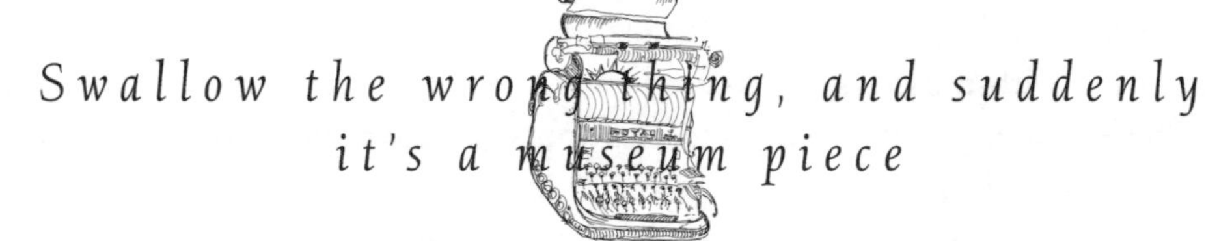

Swallow the wrong thing, and suddenly it's a museum piece

LIMA, OHIO — THE ALLEN COUNTY MUSEUM MADE my throat hurt.

Just past the iron lung and around the corner from the Buddhist shrine in the basement of this eclectic museum hangs a small display case containing, believe it or not, things people swallowed.

The things include, by my count, 90 chicken bones, 33 safety pins, six buttons, a length of hose, one-third of a tongue depressor, a nut and bolt, and assorted coins adding up to 35 cents. Not to mention thumbtacks, glass, a religious medal and the end of a skeleton key.

The case hangs, appropriately, above a water fountain.

"I'd say it's a fairly popular exhibit," said curator John Carnes. "We have a minor celebrity in there, too."

That would be Gary Moeller, the former University of Michigan football coach. As a 6-year-old Lima resident in 1947, he swallowed one of those metal whistles they used to put on the end of balloons.

How did something from Gary Moeller's esophagus end up in the Allen County Museum?

A sign above the display case explains it matter-of-factly: "Objects removed from the esophagus, bronchial tubes (lungs) and larynx of patients of Drs. Estey C. Yingling and Walter E. Yingling."

The late Estey Yingling was an ear, nose and throat doctor in Lima. If you had a bean up your nose or a penny in your gut, Yingling was the man you went to see.

In the 1930s, he began collecting the foreign objects he fished out of people. They were displayed in his office to warn others against putting things in their mouths.

His son, Walter, took up the same medical specialty and added to the display.

The younger Yingling, a center for the Ohio State University football team in the 1920s, graduated from OSU medical school, served as a flight surgeon in World War II, and was a dedicated supporter of the Allen County Museum. After retiring in 1978, he donated the collection.

Yingling died in 1996. Fortunately for history, he and his father kept careful records. Next to each item in the display are the name and age of the patient who ingested it and the date it was removed.

Were it not for this, the fact that the infant Thomas Hermiller inhaled a large wood screw in January 1950 would be lost in the mists of time.

Hermiller, of Columbus Grove, is now a 48-year-old engineer. A few years ago, he was startled when a co-worker asked him whether he ever had swallowed a wood screw.

"How did you know?" Hermiller asked. His co-worker explained that it was hanging in the museum.

"I was shocked," he said.

Hermiller, who has since dropped by to see his contribution to Allen County history, said his parents always suspected that an older brother or sister innocently put the screw in his mouth.

The screw ended up in his trachea, where it lay for several days before doctors discovered what was making young Tom sick.

The screw is one of 360 objects in the display — I am indebted to my young assistant for the day, Seth Rowe of Worthingway Middle School, for counting them.

I don't know the stories behind most of those objects, but some are not too difficult to imagine.

The two adult men who swallowed nails must have had them clamped between their teeth while hammering something.

I could almost smell the permanent wave solution and see the hair curlers when I spotted the bobby pin that slipped down the gullet of Alberta Franklin, 50, in 1958.

And perhaps Howard Hamilton, 26, was a dad ahead of his time when he swallowed the large safety pin in 1936. It certainly looks big enough to fasten a diaper.

Walter Yingling's daughter, Judy Giffin of Lima, remembers her father chuckling at the reactions of some patients after he had removed coins from them. They wanted their money back. The doctor would have to buy the coins, at unfavorable exchange rates, to add them to his collection.

The collection ends in the mid-1970s, when Walter Yingling stopped doing surgery. But he continued to worry about the risks people took with small, sharp objects, Giffin said.

"He yelled at me a hundred times about pins in my mouth while sewing."

—Saturday, January 17, 1998

Another one drives the bus; lots of barrels bite the dust

LAST WEEKEND, I KNOCKED THINGS OVER with a COTA bus.

This was part of the Celebrity Bus Driving Experience. COTA officials every once in a while invite celebrities to drive buses. But this time, they must have run out of celebrities, because they asked me.

So I found myself in the parking lot of an abandoned store in Great Western Shopping Center, with local glitterati, including the mayor of Hilliard, several Columbus civil servants and a TV weatherman.

Since my synchronized swimming ordeal last year, I've avoided all opportunities to participate in things for which I have no qualifications other than the promise of appearing inept. But, look, this was a chance to drive a bus.

As I've explained before, I love buses. I love them because they're not cars. Cars rob people of their inhibitions and cause them to commit callous acts of disregard for human life, such as shaving at 75 mph.

Bus-riding, by contrast, is sane and peaceful, at least for the passengers.

In two decades of riding, I have seen few incidents of bus rage. One time a disturbed woman did step onto the bus and throw a soft drink on a driver, but it had nothing to do with the frustrations of commuting. She was just angry at men in uniform, a state of mind that only worsened after the police were called.

In any case, I eagerly agreed to try my hand at driving.

I was told that other competitors and I would be driving 40-foot-long, 15-ton buses through an obstacle course. It was part of the same obstacle course that COTA drivers would navigate the next day in the annual Bus Rodeo.

Faced with a test of driving skill before an audience eager to witness amusing failure, I did what anyone in that situation would do: I tried to cheat. I sidled up to expert COTA driver Bob Mathews and asked him how to get through the obstacles.

Mathews, who has won the Bus Rodeo several times, said, "Make sure you don't kick anything with that big rear end." He was referring to the bus, I'm sure.

Mathews gave me detailed instruction on positioning the wheels, but I promptly forgot it all when my turn came.

I can't say I was brimming with confidence. When I boarded the vehicle, the digital sign was still flashing the name of a previous competitor, Katrina Owens of Channel 4. Longing for a pseudonym, I tried unsuccessfully to persuade my COTA guide that Katrina is my pen name.

My guide, Chris Moucha, a COTA planning official, wouldn't go along with it. Other than that, he was helpful, considering what he had to work with.

You'd think years of operating a car would make driving a bus at least somewhat familiar. But when I stepped on the gas, I felt just as confident as I would guiding a supertanker through the Suez Canal.

You're supposed to look in the side mirrors a lot when you drive a bus, but I was afraid to shift my eyes from the road ahead. So I must thank Moucha for talking me through the first obstacles. We navigated the forward serpentine challenge, the offset street test and the rear-tire clearance maneuver without crushing too many things.

No amount of talking, however, could save me in the diminishing clearance event.

The officials had set up two rows of yellow barrels to form a passageway just inches wider than the bus. My job was to drive through this narrow alley at a speed of not less than 20 mph. Moucha said that to reach the required speed in the short distance available, I should floor it. So I floored it.

The roar of the engine prevented me from hearing the barrels glancing off the right side of the bus, but out of the corner of my eye, I saw them rolling briskly down the parking lot. I had failed.

Moucha minimized my dismal performance until I pressed him for details on how many barrels were still standing.

"I think you pretty much took out the whole right row," he said. That meant the left row was still standing, I observed.

"You want to go back to pick up the spare?" he asked.

OK, so I'm not the greatest bus driver. I can still ride with the best of them.

—Tuesday, July 17, 2001

Fried-and-true: Our intrepid gourmet eats his way through the state fair and lives to write about it

I WANTED TO HAVE A FINE-DINING EXPERIENCE, so I went to the Ohio State Fair.

My goal was to put together one of those classic French multi-course meals. You know: appetizers, entree, main course with side dishes, cheese course, dessert. It would be a soup-to-nuts experience, but composed of fair food.

Why? In this era of more health-conscious diets, a man needs an elaborate excuse to eat elephant ears.

I read that the French consider champagne to be the ideal way to start a meal, but the fair doesn't sell champagne. I went with a Lemon Shake instead ($2).

It was a sassy little domestic shake, eager to please but ultimately lacking in complexity. But who wants complexity? This is the fair.

For an appetizer, I chose a new fair offering, the pork rinds at Conley's Bacon Rinds ($3.50).

"No carbs," said Ken Conley of St. Louis. "You're eating pure protein." I think he meant this as a recommendation.

Conley explained that the rinds are the dehydrated skin of pigs, stripped of all fat and meat. This left open the question of what exactly they consist of.

Nevertheless, the cooking process is impressive: Little squares of rind are dropped in 400-degree oil, whereupon they expand to about 50 times their original size. They taste like smoked air.

Next, I searched for soup. Admittedly, a steaming bowl of soup is not the first thing you think of when you think of the fair, which by law cannot start until summer heat and humidity reach shirt-drenching levels. But I wanted a soup-to-nuts experience, remember.

I finally found soup at a concession run by the Maynard Avenue United Methodist Church ("Good Food Here Since 1916.") For a dollar, I got a foam bowl of good beef vegetable soup. The stand also sells apples in their natural state — no candy coating. By fair standards, this concession is a health-food store.

Feeling much healthier, I moved on to the entree, which in a classic French meal isn't the main course but a warm-up to it. I decided to visit the Blossoming Onion stand.

A blossoming onion ($5) is made by immersing a battered onion in a trough of hot oil. It comes out golden brown, its layers fanned out like a giant fried flower. A family of four could live off one for a couple of days, especially considering that it comes with ranch dip.

Before I move on to the main course, I should explain that whenever I go somewhere,

I like to eat food characteristic of the locale.

Yes, I could have had barbecued ribs at the fair. I could have had a turkey dinner at the Ohio Food Pavilion. I could have had steak or gyros. But those things are available anywhere, anytime. I wanted something truer to the grand tradition of fair dining.

I went with the footlong coney, onions included ($2.25), at a stand near the north entrance. My side dish was a roasted ear of corn ($2) from another stand.

The corn, which I bought unbuttered because the oils from the onion still loomed large in my digestive system, was sweet and tender.

The coney I didn't eat. Let me explain what happened.

A classic French meal usually includes a cheese course, and it usually comes after the main course. The French — who can explain them? I decided to have the cheese course first.

I chose fried Swiss ($3) at Joy's Corn Dogs and Fried Cheese, a stand next to the World Famous Redwood Log House.

"Just like a corn dog, only cheese," said Gary Havens of Tucson, Ariz.

Ah, the best of both worlds.

How fried cheese is prepared should sound familiar by now. A healthy chunk of cheese is battered and dropped in the inevitable vat of hot oil. Two minutes later, they hand you a hot, weighty mass on a stick.

My advice to anyone eating cheese on a stick is to live in the moment. Because afterward, you will have second thoughts about what you have done. The stomach has a carrying capacity that fried cheese quickly exceeds.

So I passed on the coney, although I must say it looked as long as advertised.

Fair dessert choices are stunning in their variety — chocolate-covered frozen bananas, ice cream, funnel cakes, taffy, pie. But in keeping with the theme of the day, I chose an elephant ear, which is dough that is, of course, deep-fried.

Not only that, but at Gabby's Dough Factory, they also brush it with butter, dust it with sugar and top it with cherry filling for $4.50. I ate sparingly, wishing to save room for the free samples of corn nuts they were giving away near the Department of Natural Resources exhibit.

Soup to nuts. I had done it. The challenge no longer weighed heavily on my mind.

My stomach was a different story.

—Wednesday, August 8, 2001

Date with model leaves reporter nearly speechless

NOTES FROM MY DATE WITH supermodel Heidi Klum:

Wednesday, 5 p.m.: Arrive at Victoria's Secret, the fancy new underwear store at Easton Town Center. Heidi to meet me there.

Yes, that Heidi, the one who has appeared in the *Sports Illustrated* swimsuit edition. She was in town, so she called me.

OK, a friend of hers called me.

OK, a public-relations person who gets paid to call reporters called me. Hey, I can call it a date if I want to. Even my wife said I could.

5:05 p.m.: Blocklong line has formed outside store. Some women, many men. They'll probably try to call this a date with Heidi, too. Eric from Pataskala has been in line for 2 1/2 hours. Wife, a very understanding woman, is with him.

"They talked me into buying a bra," Eric says. Victoria's Secret staff told him Heidi loves to sign bras. Eric says he probably will be speechless before Heidi. Wish I had that option.

5:15 p.m.: Heidi arrives. Red, spaghetti-strap dress, stiletto heels, single-digit body fat percentage.

I'm relieved that she is fully clothed. Had feared I would have to converse with her while she was dressed in "intimate apparel."

Store festooned with huge photos of Heidi in nothing but intimate apparel. All my mental images of Heidi have her half-clothed at best. Occurs to me that I will be introduced to someone whose underwear is already imprinted on my brain.

5:18 p.m.: Heidi doing live remote with Dave Maetzold of Channel 4. They're laughing, talking. She kisses him on the cheek. Big difference between electronic and print journalism: I never get kissed during interviews.

5:19 p.m.: Struggle to think of questions to ask Heidi. Combination of TV lights, crowd, shiny fabric and perfect teeth bring Miss America pageant to mind. Consider asking what she would do to promote world peace.

5:19:30 p.m.: Tried to prepare for this interview. Really did. Research turned up only distracting things: Heidi wore live monkey on chest in a *Sports Illustrated* swimsuit edition. Some guy in New York wrote an entire play dealing with his desire to have sex with Heidi. Etc.

5:20 p.m.: Heidi's TV interview wraps up. She's turning toward me. Tall, fit, radiant. Teeth appear to have own lighting system. Have been this close to only two other celebrities: Dear Abby and that chunky guy who played Al on *Home Improvement*. Was a little intimidated by Al, so you can imagine how I feel now.

5:21 p.m.: Heidi likes my tie! Says so. I tell her I like her dress. "We are exchanging compliments," she says in German accent. Date going well.

5:21:30 p.m.: Date stops going well. Tell Heidi about Ohio State Fair tradition of sculpting celebrity likenesses in butter. Propose to start campaign to so honor her.

Don't know why I thought this was a clever thing to say. Heidi expresses perfunctory amusement but obviously has no idea what I'm talking about. Smiles and says everyone in Columbus has been very nice.

5:22 p.m.: Journalistic instincts take over. Throw tough questions at Heidi: Was it fun being in that new movie *Zoolander*? (It was.) Where are you going for dinner tonight? (Barcelona, in German Village.) Say, you're from Germany, aren't you? (She is. Aha. That explains German accent.)

5:23 p.m.: The talk turns to underwear.

What I know about underwear can be summed up in four words: Fruit of the Loom. But had read about Heidi wearing $10 million bejeweled bra.

Ask where the bra is now. Heidi says it's old news. Newer, more expensive bra to be unveiled soon. Panties, too. She quotes astounding price. Public-relations person jumps in to say price not to be revealed until later this month. Agree to stay mum because it allows me to brag that I know a Heidi underwear secret.

5:24 p.m.: Want to ask intelligent question about dress, to impress Heidi with my fashion knowledge. The color! Probably has some fancy name that Heidi would love to share with me. Mediterranean Dawn, Splash of Burgundy — something like that. I ask. Heidi looks down at the garment. "I would call it red," she says.

5:25 p.m.: Date over. I'll never forget it. Heidi will.

—Saturday, October 13, 2001

Peep show: Martha Stewart talks us into making marshmallow treats — and it's not a good thing

TO MAKE A MARSHMALLOW PEEP IS TO KNOW a heady sense of power.

Unfortunately, what I made more closely resembled a marshmallow grub.

Martha Stewart launched me on the foolish quest to produce a peep at home. She made peeps on her television show, although she called them Marshmallow Treats so as not to infringe on the trademark of Just Born Inc.

Just Born, of Bethlehem, Pa., has been making Marshmallow Peeps since 1953. The peeps have attained cult status because — well, I don't know. They're as sickeningly sweet as any marshmallow, with the added attraction of unnatural colors.

Yet, we cling to them, even as they cling to our teeth. The Internet bears an astonishing number of Peep sites, and any reference to Peeps inevitably triggers a debate as to whether they are best stale or fresh.

"I like them slightly harder than fresh," said Mikal Nolan, a member of the Ohio State University Poultry Science Club.

(Poultry scientists do not study Peeps, but I figured the subject would resonate with them.)

"Yeah, I like them," said David Latshaw, an OSU animal sciences professor who specializes in poultry. He prefers them fresh.

They don't come much fresher than the ones I made. They also don't come much uglier.

It seemed so easy when Martha did it that I, too, wanted the thrill of unlocking the secrets of a modern icon.

She heated water and sugar into a thick syrup, which she poured into an unflavored gelatin solution. Then she whipped the concoction with an electric mixer until it turned into marshmallow.

Spoon the marshmallow into a pastry bag with a No. 11 tip, she said.

Pipe the marshmallow into a bed of colored sugar to form "enchanting" peep shapes, she said.

All went well until I piped. The tip kept falling off the pastry bag.

"You're supposed to put the tip on the inside of the bag, not the outside," my wife said.

Too late now, I said.

The unstable tip made piping more difficult, but sheer inexperience was the main problem.

The first peep I piped looked like a grub.

The second vaguely resembled a duck with severe neck injuries.

I made a snake, a rock, a postmodern elephant. Where was the enchantment?

My wife said the marshmallow refused to form a peep shape because I had failed to mix it to the proper stiffness. She tried making one. It started out peeplike but then collapsed into a salamander.

"Maybe if we put eyes on it," I said.

We put two dots of food coloring on the salamander. Voila! What had looked like a sugar-encrusted amphibian now looked like a sugar-encrusted amphibian with blue eyes.

As we studied the amorphous shapes, the peeps project turned into a marshmallow Rorschach test. Does this one look more like an elephant seal or a large intestine? Does that one resemble a flattened possum or an earless mouse?

I summoned the 11-year-old, a peeps devotee, to see my handiwork. She thought I was making *Star Wars* characters.

"That one looks like Jabba the Hutt," she said.

To those who would presume to make peeps, I offer this advice: They cost only about $1 a package at the grocery store.

—Wednesday, March 27, 2002

A new car every 13 years holds plenty of thrills

DIARY OF A CAR BUYER:

March 2000: Repair shop recommends replacing old car.

My policy: If brakes work and flames don't shoot from engine, keep driving it.

March 2001: Car still running, but repair shop out of business.

March 2002: Car still running, but wife impatient.

Tell her we should wait to buy new car until George W. leaves office.

"What's George W. got to do with it?" she asks.

"I like the symmetry of replacing cars and presidents at the same time," I explain.

"Great," she says: "Most people just need money to buy a car. I have to unseat an incumbent president."

July: Odometer reading approaches that of Apollo 11.

Wife says maybe we should just look at cars.

No such thing as "just looking" at cars. Always end up buying.

I resist — must stay strong.

Aug. 25: Zero-percent financing announced on car I like.

"Maybe we should just look at cars," I say.

Aug. 30: Obtain detailed price information on Internet. Will enter dealership as informed, flinty-eyed bargainer.

10 a.m. Aug. 31: Arrive at dealership, along with everyone else within 50-mile radius. So many "lookers" that dealership is raising prices on some cars by $1,000.

"Market adjustment," sales manager explains.

"Oh, yeah? Well, I have detailed price information about how much this car really cost you," I say, flinty-eyed.

Manager looks at my information. Laughs.

11 a.m.: Arrive at second dealership.

Should have borrowed newer car: 1989 station wagon presents picture of urgent need.

We don't have to buy car today, I warn salesman. One we own is perfectly fine, despite duct tape.

Second dealership not in negotiating mood, either. Offers free carwash if we buy today. Sheesh!

Noon: Break for lunch.

Now firmly in grip of car-buying psychosis: Exorbitant sums begin to sound routine.

Hear voices telling me I want optional floor mats ($179).

2 p.m.: Arrive at third dealership, which makes great offer.

Eureka! Price is right; car is right; everything is right!

Color wrong, wife says.

Color? Color?

Color matters when buying something we might drive until 2014, she says.

2016, I correct: presidential election year.

5 p.m.: Return to first dealership, still muttering about color.

6 p.m.: Since we left a few hours ago, first dealership has sold several cars. One left. Color: black.

Have always wanted black car. Can't admit that to wife now.

6:30: Through flinty-eyed negotiation, manage to get $7 knocked off price.

Tell wife we should play hardball: walk out if dealer won't cut price more.

6:31: Dealer won't cut price more.

"Let's buy it anyway," I say.

"You just like the color," she says.

"Ridiculous," I say. "Why should I care that this sleek, sophisticated black machine will give me the mystique of Batman?"

10 p.m.: Put new car to bed in garage. Tell kids not to talk loud, because it is sleeping.

Wife rolls eyes. Says we aren't buying another one until 2024.

—Monday, September 9, 2002

Shopping trip for Valentine elicits 'ouch!' and 'ooh-la-la'

THE SHY PERSON MIGHT HESITATE to venture into certain stores in search of Valentine's Day love gear.

So I traipsed through the Garden, stepped into the Lion's Den and browsed through Wildman's Leather-n-Lace, making notes for the Shy Person.

My notes:

Basement of the Garden, 1186 N. High St., something like Ace Hardware of intimacy: You have your power tools, your lubricants, your belts and fasteners, your life-size dolls. (OK, hardware-store analogy not perfect.)

Gawk at $55 gadget made by Dr. Joel ("the most trusted man in pumping"). Comes with gauge and safety release valve, like device mechanic uses to pressure-test Toyota radiator.

Clerk Angie Fallon can put Shy Person at ease with matter-of-fact approach. Breezily displays features of best-selling Decadent Indulgence ($105) as if it were toaster.

It's no toaster. What a range of motion: vibration, rotation, gyration. And it flaps tiny pair of wings, too.

Fallon talks safety. Tells me about woman who ignored warnings against putting glass Glow Daddy in microwave; it exploded.

(Glow Daddy exactly what you think it is.)

Most expensive item: the Sybian ($1,800). Let's just call it energetic furniture: many attachments, several speeds. Highest setting could mix gallon of paint. Best seat in the house, Fallon says.

On to Wildman's Leather-n-Lace, 2264 S. Hamilton Rd.

Smaller hardware section, but how many handcuff choices does someone need?

Clerk Jackie McCoy can spot Shy Person a mile away: tends to linger uncertainly without asking questions. McCoy disarms with cheery attitude, bring-'em-along-slow sales strategy.

"We start them off with a small toy, and, if they come back, we sell them a kit."

Mr. Silky Smooth (exactly what you think it is) makes a good starter item, she says.

Recently sold a Mr. Silky to couple in their 70s. They returned and bought a kit; later phoned McCoy with profuse thanks. All in a day's work.

Wildman's costume section is extensive: nurse's outfits for those with interest in health care, school uniforms for the educationally minded.

And leather. Lots of leather. More leather than a Texas feedlot.

On to the Lion's Den, 4375 Roberts Rd., brightly lighted and colorful.

Reminds me of Blockbuster, except Blockbuster doesn't sell edible boxer shorts.

Lifelike merchandise. So lifelike, Shy Person might hesitate to touch, out of respect for personal space.

Manager Mike Lukasik — soft-spoken, businesslike, proud of orderly store — pokes and squeezes items to show Shy Person not to fear. Many toys have little "Touch me" windows.

Store also sells "make-your-own" kit ($91). Heading out of town? Use molding gel to leave memento for sweetheart; hope sweetheart doesn't leave memento in view when parents pop in for surprise visit.

Bedroom "adventure sets," toys that plug into car lighters, exotic oils, "pens" that quiver when they aren't writing, power underpants and, incredibly, a periscope — variety might shock Shy Person.

But this should be comfort: Most stores offer brown paper bags.

—Friday, February 14, 2003

Four
Modern Bewilderments

Hail to chief of unconventional speech patterns

IT ALWAYS TAKES TIME TO GET USED TO how a new president talks.

To prepare myself for George W. Bush's presidency, I have rigorously examined his approach to the English language (with the help of thc collection of Bushisms at slate.com).

The result of my research is this report, which I call "The Rhetoric and Wordplay of George W. Bush."

Those who would dismiss the president-elect's speech as confusing do not understand the unique rhetorical devices he uses to communicate.

I believe that I am the first to categorize them. Listen for these techniques when the soon-to-be president speaks:

• **Silent modification**: Bush, as he often says, trusts people — even to the extent of trusting them to supply missing modifiers. On Feb. 16, Bush said, "If you're sick and tired of the politics of cynicism and polls and principles, come and join this campaign."

He probably meant *shifting principles* or something similar. By leaving out the word, he communicates that he is a conservative, no more likely to spend tax money than he is to spend modifiers.

• **Compression**. When Bush is expressing a concept at the end of a sentence, he often severely abbreviates the thought. An example came on Aug. 30, when he said, "Well, I think if you say you're going to do something and don't do it, that's trustworthiness."

Clearly, he means that's an issue of trustworthiness, but he compresses the phrase to concentrate its impact and reinforce his image as a no-nonsense man of action.

• **The nonspecific 'it.'** On Oct. 18, Bush said, "It's your money. You paid for it."

This is a classic example of how Bush tends to unharness *it* from the preceding noun (money) and allows it to soar in search of broader definitions.

It doesn't refer to money; that would render the statement nonsensical. Bush has unexpectedly expanded the meaning of *it* to refer to government, the federal budget, and perhaps the entire universe.

In this way, Bush uses a simple two-letter word to reinforce his optimistic message about a world of limitless possibilities.

(Bush's use of the nonspecific *it* inevitably invites comparison to President Clinton's embrace of the nonspecific *is*, as in "It depends on what the definition of is is.")

• **Floating reference.** On Aug. 21, Bush said, "I don't know whether I'm going to win or not. I think I am. I do know I'm ready for the job. And, if not, that's just the way it goes."

Here, Bush is using *it* to refer not to being ready for the job but to the uncertainty of victory. This is obvious because the *if not* of the last sentence parallels the *or not* of the first sentence. *It* floats back to that reference point, skipping over the sentences in between. Once that's understood, the statement looks somewhat less alarming.

• **Fusion, Type 1.** Because a single, familiar expression is often inadequate to encompass his thoughts, Bush sometimes will join two or more in surprising ways.

On Jan. 27, he said, "I know how hard it is for you to put food on your family."

He could have stuck with the familiar *put food on the table* or the equally familiar *feed your family*. But by joining them, he adds an emotional impact lacking in either expression. Now the image is one of a family around a table. Although children are not mentioned specifically, their presence is obvious. Who else would have spaghetti sauce on their shirts?

• **Fusion, Type 2.** Bush also sometimes fuses two or more words into a single, unconventional term.

Aug. 21 is a classic example. He said, "We cannot let terrorists and rogue nations hold this nation hostile or hold our allies hostile."

This fusion of *hostile* and *hostage* undoubtedly was meant to communicate to potential adversaries that Bush's administration will monitor both their actions and their attitudes.

Likewise, his famous use of *subliminable* probably was a simultaneous reference to subliminal, sublime, interminable, nimble and limbo, a mix of concepts that may well be unprecedented in the history of spoken English.

—Tuesday, January 9, 2001

TV *asks age-old question:* Have *we won yet?*

THE WAR AGAINST TERRORISM ISN'T 2 MONTHS OLD, and already the media seems a bit impatient. What if modern TV news were reporting on past American wars?

June 6, 1944

Announcer: This is the 24-Hour News Network.

Anchor Melissa Rhinoplasty: Good evening. Roosevelt administration sources are telling 24-Hour News that Allied troops have just begun landing on the coast of Normandy in France and are being met with fierce fire from German forces. Let's go live to Frank Trenchcoat in France.

Frank: Melissa, a battle is raging nearby, and sources say Americans are suffering heavy casualties. Questions are being raised about whether this invasion, already 6 minutes old, can succeed.

(Cut to talking head, an American university professor)

Professor: Coastal invasions are risky maneuvers. The enemy will often put up stiff resistance. I would say that's what's happening in this case.

Frank: And so, how it will all turn out remains to be seen. Back to you, Melissa.

Melissa: Thank you, Frank. We have the results of an instant 24-Hour News poll that shows 73 percent of Americans are worried about the success of the invasion. That's up from 59 percent a minute ago. Now let's go live to a War Department briefing in progress.

(An American general is speaking to a room full of reporters.)

General: All I can say is that an operation is under way.

24-Hour News correspondent: Would you say the operation is a success or a failure at this point?

General (glancing at watch): Well, seeing as how we're 7 minutes into it . . .

Correspondent: When will you know whether it succeeded?

General: When we've liberated Paris, pushed through Belgium, fought our way across Germany and forced the enemy to surrender.

Correspondent: Do you expect that by the end of the day?

(The newscast breaks away from the briefing and returns to the anchor desk.)

Melissa: Now let's go to Bill Trenchcoat, who has German reaction.

Bill: Melissa, the Germans say they are succeeding beyond their wildest imagination and expect to turn back the invaders any minute now. They also say the Allies are brutal savages who are purposely targeting puppies, infants and grandmothers in Normandy. This could not be independently confirmed. Back to you, Melissa.

Anchor: When we come back: Did the Allies underestimate the enemy?

April 19, 1775

Announcer: From New York, this is ABCDEFG Nightly News with Tom Rathernot.

Tom Rathernot: Good evening. What some are calling a war for independence erupted today when gunfire broke out between Colonial militia and British troops at Lexington and Concord in Massachusetts. Let's go live to Jim Trenchcoat in Lexington.

Jim: Tom, both sides suffered casualties when the fighting began, and although the British retreated, there are fears tonight that the colonists have picked a fight with a formidable, battle-hardened enemy, and there's no telling how this day-old war might turn out.

(Cut to talking head, a university professor)

Professor: The British have highly trained troops, an enormous Navy and spiffy uniforms. The colonists are a ragtag collection of farmers with muskets.

Jim: And so, whether the colonists can succeed remains to be seen. Back to you, Tom.

Tom: Thank you, Jim. Intelligence is playing a key role in this war. With a report on that, let's go live in Bob Trenchcoat in Boston.

Bob: Tom, sources told ABCDEFG News that the Colonial intelligence network consists of one man, Paul Revere, riding a horse and shouting that the British are coming. Sources say if Revere's horse gets tired, the war effort will bog down. That may be why the American revolutionaries have recorded no battlefield successes since the skirmishes this morning.

Tom: When we come back: How much longer will this war drag on?

—Saturday, November 3, 2001

Nerves of steel get rusty when market fluctuates

TALK ABOUT A ROCK-STEADY INVESTOR. I didn't move a muscle when the Nasdaq fell off a cliff, and I barely blinked when the Dow jumped up 399 the other day.

That's because I'm paralyzed by fear.

I figure most people are. But if you ask them, they lie and say, "No, no, I'm in for the long term and can't be frightened by a little market volatility."

Right.

The baby-boom generation is the best educated in history — we know how to sound sophisticated and disciplined for public consumption. After all, we're the people who fooled the media into believing there was a "fitness craze," when in fact most of us were home on the couch, eating Twinkies.

It's likewise with the stock market. Publicly, we're maintaining a calm front. Privately, we're rending our garments and weeping over the battered bodies of our 401(k)s.

Granted, it didn't help that I had all my money in two funds: Picadilly Third World Village Bazaars and Smith-Oxley Teen-age Internet Ventures.

Until the "volatility," I was doing pretty well, especially when you consider that it has been only four or five years since I learned what the Nasdaq was. I'd heard of it before, but I thought it was a country in the Middle East.

Now? I've felt every bump on the 3,000-point slide.

Did you hear me? I said 3,000-point slide. Do you think baby boomers who thought they would retire at 50 are taking that calmly?

Think about the Nasdaq in census terms. In Ohio, a community becomes a city once it reaches 5,000 in population. So the Nasdaq, at 5,000, was a glittering metropolis. Then it fell below 2000 and became the stock-market equivalent of Mount Sterling. Even this week, with Alan Greenspan jacking it up, it's barely Sugarcreek.

This has changed my life in subtle ways. For one thing, I'm getting less exercise.

There was a time when I'd run three blocks to meet the postman because I couldn't wait to see the quarterly statement. "Any 72 percent returns in your bag for me today, Mr. Postman?" I'd ask. And he'd chuckle and hand over the envelope of joy.

Now?

When I see the mail truck, I hide under the bed.

Also, my attitudes about money have changed. In the good old days of early 2000, I believed in the reality of the numbers on the quarterly statement. Every one of those dollars existed; I could practically feel them in my wallet.

Today, I tell myself not to worry, those are just theoretical numbers, mere blots of ink that have no significance in the physical world. It doesn't matter that they're down 27.2 percent for the quarter ending March 31.

Thought barriers — those are important to a steely-eyed investor like me. What I've done these past few months is erect a series of thought barriers to prevent me from fleeing the market in favor of other investments, such as gold coins hidden under the insulation in the attic.

Barrier No. 1 was not to look at charts. Charts are the scariest thing about the stock market because they look like cardiac arrhythmia. To guard against accidentally getting a glimpse of one, I eliminated all chart-viewing. Even the Billboard 100 was off-limits.

Barrier No. 2 was not to think of the years 2001 to 2026. You need a long-term goal if you're going to be an investor, they say. So I'm pretending the next 25 years don't exist. I don't know where I'm going on vacation next year, but I have reservations at Disney World for July 2027.

When even those barriers proved insufficient to keep fears from leaking in, I went to Barrier No. 3: no numbers.

I know it sounds drastic, but every number reminded me of a Dow point-drop. Barry Bonds hits his 500th home run, and all I can think is that he's just eight away from the historic total of the '87 crash.

So I try to think only in letters, which is difficult when you're trying to balance a checkbook.

Of course, when the markets rebounded this week, all barriers came down, and I wallowed in numbers and charts for a couple of days.

Even a rock-steady investor has to cut loose sometimes.

—Saturday, April 21, 2001

Middle management: What it's really like being older than young

MIDDLE AGE ISN'T SO BAD AFTER ALL, a new study says.

At least that's what I think it says. The printing was awfully small, and I didn't have reading glasses with me.

The MacArthur Foundation interviewed thousands of middle-aged people and were told by the vast majority that they consider themselves healthy, vital and productive.

Most of the men said they had not suffered midlife crises involving desperate flings and red sports cars. Most of the women said they were navigating menopause without severe difficulty.

Gum disease, constipation, rheumatism, backaches, all the alleged maladies in middle age were professed to be relatively rare. Most people claimed to be holding together nicely.

The liars.

Not to besmirch the MacArthur Foundation, but wouldn't most of the study's respondents have been baby boomers? And aren't baby boomers famously reluctant to part with their youth? And might not that explain why all these 50-year-olds are claiming to be veritable wellsprings of vitality and productivity?

I guarantee that when the bulk of the baby-boomer generation hits old age, studies suddenly will be showing how great that is, too. After that will come the studies revealing that, contrary to popular belief, dead people are surprisingly vigorous.

The study on midlife was so upbeat that it might leave some of you wondering whether there is any difference between young adults and the middle-aged. Let me assure you there is.

How do I know? The study defines middle age as 35–65. I'm almost 45. That makes me an expert with a decade of experience.

There are many ways to distinguish the middle-aged from the young, but let's start with the important subject of getting hurt.

The chief difference between the young and the middle-aged is that young people actually have to be in motion to get hurt. The middle-aged can injure themselves just by sitting awkwardly in a chair. Getting hurt while seated — it may be the signature condition of midlife.

Speaking of sitting, young people do it without comment. When someone my age sits, they like to issue some kind of statement, usually a groan. When they get up, they groan again. This is thought to be a way to alert any laggard joints that they should unlock themselves immediately and get moving.

As for sports, it's obvious that the young have more athletic prowess. The middle-aged, however, have the edge in decision-making and judgment. Pit the agile-but-impulsive young against a team of slow-but-savvy middle-agers in a demanding sport such as touch football, and the results can be surprising. The young will win 49–0. But it will be the middle-aged who have the foresight and taste to bring along a good brand of imported beer.

As for intellectual ability, young people have a slight edge. The young, for example, read fast. The middle-aged read equally fast, but their preparation time is longer. They have to remove their glasses, hold the page at arm's length, squint, mutter, squint again and go in search of a lamp with a 500-watt bulb.

Memory is another key area of difference. In a study conducted at a neurological research clinic, young adults presented with long lists of words were able to remember most of them after hearing the list repeated just twice. The middle-aged fared much worse, in part because 47 percent of them completely forgot they had agreed to participate in the study and went to see their liposuctionists instead.

Dietary habits also differ. The young can spend the night drinking, top it off with a 2 a.m. meal at White Castle, go home and sleep like babies. The middle-aged can spend the night drinking, top it off with 2 a.m. White Castles, go home and feel like they are having a baby.

The young also need less sleep. The middle-aged function best on a full night's rest, although some can get by on 10 or 11 hours of sleep without a great loss of productivity.

These aren't all the differences, of course. I'd mention more, but if I sit in this chair any longer, I might get hurt.

—Saturday, February 20, 1999

Guns are only tools— like grenades and land mines

AS YOU KNOW, THE GUN CROWD HAS SOME sound arguments for why we need lots of weapons in this nation. As the owner of a caulking gun and two Nerf pistols, I think it's time I do my part to support their views.

This, after all, is the season when our thoughts turn to guns. Spring blooms and troubled teen-agers head off to their proms, their sidearms color-coordinated with their tuxedos.

Though we're now averaging a school shooting rampage every two to three weeks,

I want to assure you that it has nothing to do with guns.

Guns are simply inanimate objects poised to send high-speed projectiles crashing through your liver. To suggest that they somehow play a role in mayhem is a far-fetched argument indeed.

People who want to control guns conveniently ignore the Second Amendment to the Constitution, written by the Founding Fathers, who had perfect foresight. They foresaw that high-powered weapons would be developed in the 20th century. And they foresaw that we would need unfettered access to them because Americans in 1998 would be under constant threat from the huge buffalo herds roaming the plains.

Therefore, any call for gun control is un-American. Not only that, it gets Charlton Heston angry, and he might retaliate by making another movie.

Heston is a member of the National Rifle Association, an organization that is sort of like the Tobacco Institute of guns. The NRA's mission is to preserve domestic tranquillity by buying congressmen and forcing them to attend endless rounds of cocktail parties.

Much has been written lately about automatic weapons. I want to stress that automatic weapons are not the problem. Or maybe it's semiautomatic weapons that aren't the problem. I can't keep them straight beyond knowing that both are capable of reducing someone to a bloody pulp in three seconds.

And, really, there's no sense getting hung up on fine distinctions between these weapons. It's probably just as well to concentrate on finding available cover.

I used to fear firearms. I mean, I didn't even like to be in the same room when Martha Stewart was wielding a hot glue gun.

But the gun people have explained to me that a gun is nothing but a tool, like a rake or a hoe. Used properly, such as to shoot weeds in the garden, it is perfectly harmless. This has reassured me.

You may still be wondering, however, why an average unwounded citizen like me is representing the arguments of the gun crowd.

Because if average citizens don't stand up for the unrestricted right to bear anything that goes boom, this country could be taken over by slick politicians and the giant corporations that fund their campaigns. Soon the IRS would be requiring us to fill out incomprehensible tax forms, the White House would be occupied by a gang of hedonists, and not even the freeways would be safe from incivility and violent outbursts.

And what about threats from beyond our borders?

Right now, all around the country, gun-owning patriots are snoring in their recliners. But they're ready to spring up at a moment's notice, and — if they don't pull a muscle — grab rifles and take to their tree stands. There, the patriots will plink away at massive columns of enemy tanks and helicopter gunships as they pass by on their way to seizing the state capitol.

For these reasons, we must be ever-vigilant against schemes to take our weapons away. Beware of tricks such as a guns-for-Viagra exchange. Beware of those who claim all they want to do is restrict armor-piercing bullets or rifles designed for combat infantry use.

They won't stop there. They'll chip away until they've taken even our salad shooters.

Do you know where even modest gun control could leave us? With unpierced tin cans piling up in the nation's trash stream, our shrubs nibbled to the ground by deer, and thousands of sidewalks unadorned by chalk body outlines.

The millions of us who love gun freedom should march on the U.S. Capitol to make our point. Allow two weeks. It will take that long for all of us to pass through the metal detectors.

—Tuesday, May 5, 1998

Resistance seems futile when spam fills mailbox

MAYBE IT WAS THE WINTER DOLDRUMS. Maybe it was a midlife crisis.

In a moment of weakness, I accepted all the e-mail offers — the spam — I had been receiving for years.

Ludmilla, the lonely Russian who beckons electronically? I invited her to come live with me.

"You did what?" my wife said.

"Honey, look at this e-mail: There are many other Russian women like her seeking American companionship. Let's reach out to people of other lands."

Ludmilla showed up two weeks later, a sullen, bottle-blond chain smoker who speaks little English.

We put her in the guest bedroom, already crowded with laser-printer toner and septic-tank cleaner.

Day after day, e-mail messages had urged me to take care of my septic-tank problem: They convinced me I had a problem. They convinced me I had a septic tank.

"You spent $300 to clean something we don't have?" my wife said.

"Soon, $300 will be like 3 cents to us," I said mysteriously.

I hadn't told her yet about the e-mail from the widow of the Nigerian ruler, arriving like a gift from heaven just as I was facing thousands in Internet entertainment bills.

(Those "See Britney Nude" offers? What they don't tell you is that the access charge is $109 a minute and that the nude in question is Britney La Boomba, an "entertainer" with cellulite.)

The Nigerian widow sought an "urgent business relationship" that would help her spring $35 million from the foreign bank accounts of her late husband.

I would get 30 percent.

Why did she choose me? Educational credentials, I'm sure, The solicitation came just days after I bought a doctorate in finance.

"We can assist with diplomas from prestigious, non-accredited universities based on your present knowledge and life experience," the e-mail had said. "No one turned down."

I also bought a bachelor's degree for Ludmilla, thinking it would cheer her up.

She remained sullen.

"She thought she was coming here to marry you," my wife said. "I want that woman out of the house."

"Maybe if she had a job, something to keep her busy — a home-based business, perhaps."

I had already spent $500 on instructional materials that were supposed to teach me how to unlock the secret of "the most profitable marketing technique ever created — BULK E-MAIL."

I envisioned Ludmilla happily employed, making huge profits for me.

Alas, her language deficiencies thwarted the plan.

She tried to market miracle-diet pills with this message: "Eat all borscht you want, lose much kilograms."

And so I arrive at the part that most embarrasses me: Yes, I fell for the, um, body-enhancement offers.

Enlargement, enlargement, enlargement — the messages came every day. Feeling inadequate, I sent $200 to a company promising enlargement.

What did I get? An 11-by-17 photograph of a well-endowed male. An enlargement, see?

When she saw the photo, Ludmilla began packing for Moscow — but not before she slapped me.

I gave her some going-away gifts: an international driver's license, a cable descrambler and a certificate that declares her an ordained minister.

She spit on them.

I hope the Nigerian widow is easier to please.

—Friday, January 10, 2003

Hotshots know how to conceal in high style

WITH THE CONCEALED-WEAPONS LAW PASSED, Ohioans soon will be able to incorporate stylish guns into their wardrobes.

Talk about to-die-for accessories!

Don't let the challenge of concealing a bulky firearm concern you. If you use good taste, you can pack heat and look hot, style experts say. Then you'll be a victim of neither crime nor fashion.

Here are answers to some common questions about guns and fashion:

Q: I'm getting married in June, and I need advice on wedding weaponry. My mother says I should incorporate a floral-patterned revolver into the bridal bouquet. But that's too cutesy for me. Any tips on how to conceal a weapon in my gown?

A: Wedding planners preach this axiom: Pick the gown, then the gun.

Are you wearing one of those slinky sheaths? Then you don't want the bulge of a .44-caliber revolver marring the slim silhouette.

Instead, consider strapping a dainty derringer to your thigh. Just make sure that the groom doesn't try to remove the garter from the wrong leg. He could get quite a surprise!

Your options are greater if the gown has a long train. Some of the new, personal howitzers are light and could easily be towed by a bride who wants to put all those yards of satin to good use. Besides, wouldn't an artillery salute really pep up the chicken dance?

Q: Can I wear white pearl-handled revolvers after Labor Day?

A: If they're concealed, sure. And even if they're not concealed, who's going to argue with you?

Q: Is it bad taste to let shoulder-holster straps show?

A: It used to be considered a no-no, but not anymore. Now holster straps are seen as a sexy accessory, just like bra straps.

Be sure to pick the right holster strap for the right occasion:

Delicate spaghetti holster straps go well with evening gowns. Thick, studded black straps look great at motorcycle-gang meetings.

Hitting the beach? Several arms makers are bringing out bright, playful beach holsters (look for the brand names Tutti-Shooti and Sun 'N' Gun) that will add pizazz to any bikini.

But know the risks of packing at the shore: The shoulder holster will give you a third tan line.

Q: Does the layered look work with concealed carry?

A: Certainly. Just don't overdo the layering. By the time you unzip your coat, unbutton your sweater, open up your shirt and reach into your long underwear to pull a gun, you might forget whom you intended to shoot.

Q: I want to look professional at work. What kind of weapon should I conceal?

A: Good question. A chunky black revolver is great for a casual day on the town, but at the office you want something a little more sophisticated and professional. A slim, silver-plated firearm with built-in picture phone and personal digital assistant will say that you mean business.

Q: What kind of lingerie goes well with a snub-nosed .38 on a romantic evening?

A: Try a bullet bra from Victoria's Secret Weapons Cache.

Q: Clothes are getting more revealing at the same time that laws about carrying weapons are loosening. What happens when clothes are too skimpy to conceal weapons?

A: Then the weapons themselves will be the clothes! Crisscrossing bandoliers and a strategically placed pistol will provide barely-there coverage that will look smashing and give new meaning to the title Naked Gun.

—Friday, January 16, 2004

Putting SUVs to good use: Big ugly gas guzzlers will become public art

Field of SUVs, THE CONTROVERSIAL PUBLIC ART project that uses abandoned sport-utility vehicles as sculpture, will officially be unveiled today in Dublin.

The $500,000 installation, an effort to recycle SUVs made unaffordable by high fuel prices, has been hailed as a witty comment on America's car worship and derided as a waste of tax money.

When the enormous tent hiding the project is removed today, the public will see Jeep Cherokees, Ford Explorers, Lincoln Navigators and other sport-utility vehicles artfully stacked in a field in an arrangement reminiscent of Stonehenge.

"Some people already are calling it Jeep-henge," said Dublin Mayor James Joyce.

The project is expected to put a small dent in the growing problem of what to do with the mammoth vehicles that suburbanites are abandoning along roads because they can't afford to refill them with gas.

"Land Cruisers, Land Rovers — you name it, we're finding them," Joyce said. "In every case, the needle is on empty. I saw an entire youth soccer team pushing an Isuzu Trooper on 161 the other day."

After several weeks of controversy, officials are hoping for a positive response when the art is unveiled at 1 p.m. in a field south of the Dublin Indoor Lacrosse and Rodeo Arena.

The supervising artist, an enigmatic San Franciscan who goes by the single name Solo, insisted that the installation be erected undercover to add a sense of drama to the project.

"The drama. Very important big fun. Make people weep with joy," said Solo, whose vaguely foreign speech patterns have become the subject of much speculation since *The New York Times* revealed that he was born in New Jersey.

To erect *Field of SUVs* without attracting undue attention, participating suburbs quietly collected 67 abandoned vehicles, 40 of which were used in the installation. The 27 others were parked in scattered locations and set on fire to distract TV news crews.

This brought protests from environmentalists, but the acerbic Solo has been unapologetic. He said studies show that burning an SUV causes less pollution than driving one.

Field of SUVs is a joint project of Dublin, Worthington, Westerville, Gahanna, Powell, Hilliard, New Albany, Bexley and Upper Arlington, with each contributing about $55,000 and several SUVs. Officials say the project, which can be expanded to accommodate up to 200 SUVs, will pay for itself in reduced vehicle-disposal costs.

The idea grew out of meetings held by the Mid-Ohio Regional Planning Commission in January to discuss what to do about derelict SUVs that were piling up on streets and highways.

One early proposal — giving the SUVs to the homeless — made national news in February and was the subject of jokes by Jay Leno on *The Tonight Show*.

Westerville Councilman Cal DeSac, who calls the art project "a glorified junkyard," continues to insist that the homeless proposal was a more sensible alternative.

"The fact remains that you could easily house a family of four in a Chevy Suburban. But the media twisted the idea all around. One of those tabloid TV shows reported that the homeless would also get vanity plates. After that, we never had a chance."

Even after agreeing to the art project, officials nearly had to scrap it amid bickering over the location.

Worthington was an early favorite as host site, but its city council said the display would be incompatible with the suburb's Colonial look. Solo offered to repaint each SUV in a red brick pattern with shutters, but the council said no.

New Albany officials then agreed to take *Field of SUVs* but reneged in the face of outrage from homeowners. Solo unsuccessfully tried to mollify them by saying he would place the entire display inside a 17-car garage to match those in the neighborhood.

Just when it looked as if no suburb would accept the project, Dublin stepped forward. Already home to *Field of Corn,* the giant concrete corn sculptures at Frantz and Rings roads, Dublin agreed to host *Field of SUVs* provided the other suburbs pay for a fence around it.

"We don't want anyone suing us because they got hurt on the SUVs," Mayor Joyce said. "We also need a way to keep Fred Ricart from sneaking in to put advertising on them."

—Saturday, April 1, 2000

Dream team retools our yule

'Twas the night before Christmas, with us thanking heaven
that we'd narrowly avoided Chapter 11.
The credit cards were iced in the basement with care
in hopes they'd cool off just in time for next year.
The kids had been wrestled and dragged into bed,
but a long night of drudgery still lay ahead.
Electronic playthings invented by geeks
awaited assembly, with instructions in Greek.
And Ma near exhaustion, her head badly fogged,

said, “The ham’s not yet baked and the egg’s not been nogged!
Our 20-pound turkey’s a half-frozen bird!
We’ve had three hours sleep since December the third!”
“Don’t remind me,” I said, as I blinked back a tear,
then what to our wondering eyes should appear
but a self-assured woman all dressed like an elf,
whom I realized to be Martha Stewart herself.
And behind her a creature with green skin and hair:
“The Grinch,” hissed my wife, “now there’s an odd pair.”
Scrooge entered then, and by then we were weeping;
“What’s next?” I exclaimed, “Eleven lords leaping?”
Then Martha commanded her team to begin
getting our house into order again.
In a daze I heard Martha say, “Turkey must go,
truffle-stuffed pheasant’s much nicer, you know.
And I’ll redeck their halls to my own perfect taste,
now where’s my hot glue gun — there’s no time to waste.”
Scrooge loaned us money with no interest at all
(lest the three ghosts of Christmas revoke his parole).
The Grinch muttered curses but worked through the night
until every last toy was assembled just right.
“Downsized me into this job,” he said, sour,
“The Whos got a teen Grinch for five bucks an hour.”
At 7 a.m. they declared our house done,
and the three of them vanished as quick as they’d come.
And then I looked around and said, “Ma, can this be?
Our house looks like something on HGTV.”
She sighed, “It’s a miracle, but not one so great
that the kids can’t undo it by a quarter to eight.”

—Friday, December 6, 1996

On spare tires and the auto industry

THIS IS THE STORY OF THE DETROIT PROJECT AND HOW IT unleashed the awesome power of fat on America. It began as an attempt by the car companies to enlarge people so they would want to buy bigger, more expensive vehicles. Soon it was out of control, and now, two decades later, an entire nation lives under the threat of fat.

Before 1960, fat was just another element of nature, like air, water, parsley, ear lobes and dachshunds. But in that year, a scientist named Albert Greenfield became obsessed with the idea that fat had unique properties that, if harnessed, could alter the course of history. Greenfield was a scholar of eclectic tastes; he had studied the writings of Einstein, Henry Ford and Betty Crocker. Influenced by them, Greenfield came to believe that the future industrial strength of the United States could be assured through weight gain. In 1960, he proposed the theory that would spark a revolution: People will expand to fit the size of their vehicles.

The automakers were intrigued. Together, General Motors, Ford and Chrysler launched a research effort to see if fat could indeed be put to work for America. They called it the Detroit Project. It brought together physicists, chemists, automotive engineers, economists, nutritionists, pastry chefs, short-order cooks and bartenders in a bold scientific endeavor.

"Everything was top secret," remembers one researcher. "Mr. Greenfield told us that the project was a matter of national security. He said if we didn't succeed, the Germans and Japanese might, and we'd be invaded by a horde of nine-passenger Toyotas."

Enthusiasm ran high, but the early experiments with animals were often frustrating.

"We had all these little scale-model cars with real engines," a researcher recalls. "And we had white rats. The idea was that every time a rat ate all his food pellets, you rewarded him with a larger car. It worked at first; but once we got up to Pontiac Catalinas, the rats quit eating. The Pontiacs were just the right size for two rats to mate in, and that's all they cared about. We tried teaching them to drive, but it was a bust, even with automatic transmissions."

The automakers made a controversial decision: They would switch to human experiments. Twenty men who drove compact cars and wore suits no larger than 38 medium were kidnapped and taken to an isolated testing area in the New Mexico desert. The success of these experiments depended on many factors, but food was a key element. The chefs of the Detroit Project had been brought to New Mexico with one mission — to prepare the greatest repast of fattening foods ever assembled. The menu for just one dinner reveals the scope of their accomplishment: whole roast suckling pig in cream sauce, side-of-beef Wellington and, for dessert, a chocolate moose.

"I remember, I took a quail," says a chef who was there, "and I stuffed it into a Cornish game hen, which I stuffed into a chicken. Then I put the chicken inside a duck, the duck inside a goose and the goose inside a turkey. A six-bird entree. And you know what? Mr. Greenfield reprimanded me for not squeezing a capon in there somewhere. That's what kind of pressure we were under."

But the demanding Greenfield was seeing results. The human guinea pigs, presented with cars of increasing size, responded by eating and gaining weight, knowing that even bigger cars awaited them. Greenfield's hypothesis was being proved. The time had arrived to bring out the Cadillacs.

"I'll never forget C-Day," says a project scientist. "By this time, the guinea pigs all weighed 300 pounds. They were so fat, we just dressed them in loincloths. The fat guys were blindfolded at one end of a mile-long test track. At the other end, we parked the Cadillacs. In the middle was the food. We were all in concrete bunkers, because no one knew what might happen."

Shortly after 10 a.m. on the morning of July 4, 1961, the blindfolds were removed, and 20 obese men in loincloths began walking toward the cars that shimmered in the distance. When it became clear that the cars were Cadillacs, the obese men started running.

"There was this terrible rumbling," another scientist remembers, "like the whole Earth was shaking. We were awestruck and very frightened. The energy being released was incredible. Just when we thought the bunkers might collapse, the subjects reached the food tables, sat down and started eating. You could see their waistlines mushrooming before your eyes. That was the world's first look at the Cadillac Effect."

Word was flashed back to Detroit, and within weeks, the production of big cars was quadrupled. In response, America ate. The car-fat chain reaction had been set off, and a new age was born. The influence of fat would come to pervade virtually every facet of American culture in the 1960s:

Foods such as Hamburger Helper and Cool Whip were invented and consumed in vast quantities. Television sets from coast to coast blared the message "There's always room for Jell-O." It was the Golden Age of Red Meat.

Jackie Gleason soared to the top of the television ratings, and a flabby cartoon caveman named Flintstone became an American folk hero.

Fast-food restaurants flourished as car-crazed Americans rushed to cram calories so they could move up to the next-size model.

A decade of frenzied eating and unrestrained buying followed. Believing that nothing could stop them now, the automakers began planning a new generation of mammoth vehicles. GM rented a huge warehouse and began building a prototype of the Pontiac Queen Mary — it was large enough to be locked through the Panama Canal. On the drawing boards at Ford was the Ford Twolane, a car that used the entire width of an interstate highway. It was a heady, creative time in Detroit, but it would not last. In 1973, disaster struck in the form of the Arab oil embargo.

"You want to talk about a mess," says a retired auto executive. "We had fattened up the country, big cars were selling like hotcakes, and all of a sudden, there's no gas to put in them. We had to do something fast."

The obvious answer was to produce smaller cars, but the burning question was how corpulent consumers were going to fit inside them. In their desperation, the automakers resorted to two solutions that haunt the nation to this day: polyester and disco.

"We figured that people in polyester pants could slide under the steering wheels of small cars easier," says the retired executive. "And then somebody got the bright idea that if you could make people sweat in polyester, they'd really slim down fast. That's where disco came in. OK, so polyester was a blight on the landscape; it really wasn't that harmful. But disco — I'll never forgive myself for that one."

Into this picture of gloom stepped a name from the past. Albert Greenfield. Now aging and guilt-ridden, Greenfield was haunted by what his theory had caused. The victims of double chin, midriff bulge and dimpled thigh were everywhere, and their chubby faces flashed before Greenfield's eyes when he tried to sleep at night. His own son, only 28 years old, had developed saddle bags. And the large cars he had placed so much faith in were now virtually useless.

Unable to live with himself, Greenfield set to work in search of a solution. He toiled around the clock, forsaking sleep and food, until the pace finally caught up with him and he was hospitalized. Even then, he insisted that a Mercury Marquis be stationed at his bedside so the work could continue. His family begged him to stop and rest, but Greenfield ignored them. Finally, minutes from death, the starved, exhausted Greenfield found the answer. He scribbled out the theory and then died. His scrawled message said: "People will also shrink to fit the size of their vehicles."

The automakers rejoiced. Hastily, they re-tooled the factories and poured forth a stream of four-cylinder cars. In a pattern reminiscent of the '60s, Detroit cranked out the small cars — and jogging, Jazzercise and *Jane Fonda's Workout* book soon followed. The downsizing of America had begun.

Of course, when the oil embargo faded from memory and gas prices fell, the process reversed itself again. Eager to sell high-profit SUVs, the automakers coaxed fast-food restaurants into the super-sizing craze. Once again, it was no coincidence that Americans and their cars grew plump at the same time.

And so the car companies survive. But their efforts to fatten, then slim, then re-fatten the population have left a nation divided. On one side are the unhappy obese, struggling constantly to lose the weight the automakers put on them. On the other side are the fanatically fit, exercising endlessly in a grim effort to look like their sleek vehicles.

Will America conquer its obesity problem? Look to the next Detroit Auto Show for clues.

—Sunday, December 8, 1985
Updated, 2004

Five

Not-So-Deep Thoughts

Other designs deserve some credit for human existence

THE CONTROVERSY ABOUT THE ORIGIN OF LIFE has focused on the challenge to evolution by proponents of intelligent design.

But why ignore other theories?

Intelligent design — the proposition that a purposeful creator was responsible for human existence — isn't the only alternative.

Today, So To Speak examines lesser-known hypotheses about how we got here:

• **Interior design**. The planet remained a lifeless rock until the Master Decorator got bored with earth tones, adding more-vibrant colors and a water feature.

The continents were originally joined, but the Decorator separated them because well-defined spaces are more visually pleasing.

Also, the Decorator feared that the zebras clashed with the toucans.

• **Residential design**. To hold down costs, the Grand Builder sprinkled the universe with identical planets on identical galactic cul-de-sacs.

Some planets have thrived and others have failed, depending on the competence of the subcontractors.

A few aging planets have tried to revive themselves with aluminum siding.

• **Unintelligent design**. Life arose on Earth when a Supreme Television Network Programmer started a reality show that got out of hand.

The cast has become so unruly that even Fox has doubts about airing the series.

• **Postmodern design**. This theory holds that all humans belong to an elaborate performance-art installation mounted by the Ultimate Artist, who received a really big grant.

The Artist also put the rings around Saturn. A plan to cover Jupiter in chocolate was scrapped for lack of funding.

• **Industrial design**. The Great Magnate proposed a human-populated planet, which he said would create jobs and boost tourism in the universe.

The Not-in-My-Galaxy movement objected. Humans, it said, would pollute the environment and keep solar systems awake at night with their noisy wars.

Eventually, the two sides compromised: The Magnate agreed to build one planet separated by a buffer zone of 600 billion light-years from more civilized portions of the universe.

He also agreed to add landscaping.

• **Fashion design**. Humans were created because the Almighty Designer couldn't get the orangutans to wear stiletto heels.

The humans started as large, hairy, brutish creatures but evolved after Joan Rivers made fun of how they looked in strapless gowns.

• **Ergonomic design**. Earth began as a harsh place of erupting volcanoes, poisonous gases and extreme temperatures.

Then a Supreme Consultant was hired to improve air quality, install a new heating and cooling system, and make animals more comfortable to sit on.

The planet became more productive, although humans still complained incessantly about the weather.

• **Software design**. The globe was invented by a committee whose technologically gifted members hoped to spend years perfecting it.

Unfortunately, the Chief Executive ordered it into production before all the bugs were worked out.

Earth 2.0 might be better.

• **Classic design**. Aiming for durability, the Supreme Sculptor originally made people out of marble.

The Sculptor switched to flesh because Venus de Milo kept losing her arms.

• **Automotive design**. On the drawing board, humans had indestructible teeth and long-lasting hair.

Unfortunately, production costs proved too high.

The Omnipotent Manufacturer substituted cheaper materials but gave us brains so we could invent dentures and toupees.

—Thursday, March 28, 2002

Whenever states get together, fighting words fly

ON A COLD WINTER DAY, OHIO WALKED into the Union Tavern ("Where the 50 States Meet") and nodded to Kentucky, who was nursing a bourbon and studying a racing sheet.

At a nearby table, in sophisticated black, New York was eating a bagel and reading the arts section of the *Times*. Meanwhile, Washington was sipping a latte and checking e-mail on a laptop.

Ohio always felt a little uncomfortable around the coastal states.

Suddenly, New York glanced at Ohio.

"Hey, Iowa, how's it going?" New York said.

"It's Ohio," said Ohio, feeling frumpy in a flannel shirt.

"Iowa, Ohio — same vowel-consonant ratio," New York cracked.

Ignoring the remark, Ohio chose a stool at the bar next to Michigan.

"What're you drinking?" the bartender asked.

"Flaming Cuyahoga," Ohio said.

As he mixed the drink, the bartender made small talk: "How ya doing, Ohio?"

"Well," Ohio said, "I turn 199 this year, but sometimes I feel about 250."

"Ah, you look good for somebody born in 1803," the bartender said.

"Got a touch of unemployment," Ohio said. "And my legislature is acting up again."

The bartender sighed.

"Seems like every state I know has legislature trouble — inflammation, rot, what have you."

Just then, Florida entered wearing shorts, a tropical-print shirt and sunglasses.

Ohio and Michigan stared.

"Dressed like that in the middle of winter," muttered Michigan, who in the winter wears gloves and a hat, even indoors.

Both states secretly welcomed the Florida distraction: In the absence of a Sun Belt state to unite them in mutual insecurity, they get testy with each other.

Michigan has bigger Great Lakes and never lets Ohio forget it.

Just then, California walked by clutching a bottle of cabernet and flashing a brilliant smile.

"Iowa!" California said. "How's it going?"

"It's Ohio," Ohio said through clenched teeth.

California slid into the sunny booth where Florida sat.

Soon a crowd gathered; everybody seemed to gravitate toward those two.

California has tremors and respiratory problems, and Florida is troubled by a stormy past, but they still manage to project a glamour that eludes the Midwestern states.

Suddenly, the front entrance opened, and Ohio felt a blast of chilly wind. In the doorway stood the Dakota twins.

The Dakotas are rangy, rugged, unpretentious types; South Dakota, with chiseled features, is the better-known.

North Dakota has a frosty air that puts off some people.

In their presence, Ohio always felt a little more popular by comparison.

Ohio exchanged greetings with the twins, then turned back to Michigan.

It was time to make a little trouble.

"Summer'll be here before you know it," Ohio said.

Michigan heard the comment for the challenge it was.

"Yep," Michigan said, "and I've got Grand Traverse Bay."

"And I've got Put-in-Bay," Ohio replied sharply.

Michigan arched an eyebrow and responded with an emphatic "Upper Peninsula."

"Amish country," Ohio said fiercely.

"Greenfield Village!"

"Cedar Point!"

Everyone else enjoyed the Ohio-Michigan arguments: Nevada looked away from a poker machine in the corner. Texas stopped bragging to listen. Even Alaska, usually a little distant, took an interest.

The argument ended the way it usually does at this time of year: Ohio said, "Kelleys

Island," and Michigan countered with "Mackinac." Both were reminded of fishing, then, because of the season, ice fishing.

"Hey, want to go see if the ice is thick enough?" Michigan asked.

"Why not?" Ohio said.

They rose to walk out, nearly knocking over Rhode Island, who always seemed to be underfoot.

Deeply tanned Arizona, carrying golf clubs, was arriving as Ohio and Michigan were leaving.

"Where are you two going?" Arizona asked.

"Believe me," Ohio said, "you wouldn't understand."

—Tuesday, January 1, 2002

His numbers would make even Enron embarrassed

TODAY I ANNOUNCE THAT I INTEND to restate my earnings and accomplishments for the past 12 quarters.

I do so with full confidence that the people emotionally invested in me, not to mention my creditors, will retain unshaken faith in my integrity.

A routine internal review of my finances reveals that I did not actually win a $78 million lottery jackpot in the fourth quarter of 2000.

Instead, I won $78 at bingo and inadvertently added six zeros when filling out the income portion of a credit-card application.

This might not conform to generally accepted accounting principles. Therefore, I have ordered that the earnings statement be revised.

The accounting safeguards that brought the problem to light have worked exactly as intended and should serve to reinforce investor confidence in me.

I thank my credit-card company for the $6 million credit limit it granted based on my lottery-earnings statement. I plan to make the monthly minimum payment until the debt's retirement or the year 2350, whichever comes first.

The auditors also have recommended that I clarify the business losses I claimed on my 2001 tax return in an effort to offset the $78 million in extra income.

While the garage sale I had in the second quarter of 2001 did not turn out as successfully as I'd hoped, the losses did not total $102 million. I have directed that the figure be revised to $9.37.

I am confident that this corrective action demonstrates a continuing commitment to sound business practices — and will impress the judge at my sentencing hearing.

As for the matter of my winning the Academy Award for best supporting actress: When inconsistencies were detected in my earnings statement, I thought it only prudent to order a comprehensive review of my resume.

Many facts were found to be in order, but the auditors did request clarification regarding the Oscar.

After a thorough and rigorous review, several conclusions were reached:

I am not an actress.

The only award I received in 2001 was bestowed on Customer Appreciation Day, when Pizza Hut awarded a free soft drink with the order of a large pepperoni pizza.

Also, I am not the poet laureate of the United States.

I am confident that this prompt correction of my record again demonstrates a solid commitment to trustworthy conduct.

Therefore, with such adjustments behind me, I announce restructuring plans aimed at positioning my household for a stable future:

• In the third quarter of 2002, I will divest myself of the dog.

The dog, while viable, has not met productivity forecasts. And she keeps throwing up on the carpet.

My goal of finishing the third quarter in the black remains within reach, as long as I sell the dog for a minimum of $200,000.

• I will redouble efforts to find pockets of profitability within the operation.

Already, I have found $1.56 under the sofa cushions.

I expect to search beneath the car seats soon.

• For the rest of the third quarter, I will pay creditors in arcade tokens.

With a renewed focus on fiscal responsibility and ethical behavior, I am confident I will emerge as strong and respected as any American corporation in the current business climate.

—Friday, July 26, 2002

Edgy maven seeks full closure on most-hated words

I LOVE WORDS, BUT NOT ALL WORDS.

There's little logic to the words on my hate list. I just hate them, that's all.

I can't stand the word *gubernatorial*, for example. I avoid writing it at all costs, not that the costs are that high, since I rarely write about governors.

My distaste for *gubernatorial* isn't political. I just think it's an ugly word, a 13-letter monstrosity.

"Write the way people talk" is one of my favorite bits of writing advice. Can you imagine leaning over the back fence with your neighbor and saying the word gubernatorial? I can't, although I may not be the best person to ask, since I have no back fence.

When I see *gubernatorial*, I always want to spell it *goobernatorial*. In some states, that's probably the most appropriate spelling, but let's not get sidetracked by politics.

I'll reveal the rest of my most-hated-words list if you promise not to remember it. The day will undoubtedly come when I'll need to use one of these words, and I don't want to be accused of hypocrisy.

For example, if a governor ever has gallbladder surgery, I'll probably want to write,

"Doctors removed the gubernatorial gallbladder today." Otherwise, I can't see using the word.

My hate for some words is based on sound, not meaning. Thus, the word *maven* is on my list.

Maven is a synonym for *expert*, but it sounds derogatory to me. It makes me think of large, black birds squawking lines from Edgar Allan Poe. Will I use it? Nevermore.

Not even a flock of mavens could convince me that I should use the word *eponymous*. It's an adjective signifying that something, such as a business or work of art, has the same name as its creator.

Thus, Martha Stewart's magazine is eponymous because it has Martha Stewart as part of its name. I suppose I could save a few words by writing, "Martha Stewart's eponymous magazine" rather than "Martha Stewart's magazine, *Martha Stewart Living*." But I would feel like a showoff, a role that I think is best left to Martha herself.

I hate the word *kudos* because no matter how you use it in a sentence it looks wrong. *Kudos* is (or should that be *are*?) a word that always looks plural, even though it's always singular. There's no such thing as a kudo, which is too bad, since some people deserve no more than that.

Speaking of praise, I don't much like the word *encomium*. It's too highbrow, and there are too many n's and m's to trip over. It means "formal expression of high praise." High kudos, in other words. It's very poetic, but I can't write it.

Closure wasn't a bad word until it became strained from overuse during the Oklahoma City bombing trial, when the media said everyone was seeking it. It's psychobabble (I love that word). I've pretty much abandoned *closure*, except when I need it for irony or in crossword puzzles.

I've also marginalized *edgy*. It's fashionable now to use *edgy* to describe something that's out on the edge and a bit dangerous. Edgy drama, edgy music, that sort of thing. It's become a cliche. Shouldn't a word that signifies being out on the edge be out on the edge itself?

Gaming? Ugh. I can't imagine anyone saying, "Hey, Louie, let's go gaming in Vegas." It's much too prissy a word for gambling.

Gaming is obviously the construction of a public relations department somewhere, all the more reason never to use it. I'd like to think that the mob, which invented Las Vegas and doesn't have a public-relations department, would whack anyone who went to Vegas and said *gaming* in public.

Cyberspace is a word I hate. This is partly because of overuse and partly because of pretentiousness. What an overstated way to describe staring into a computer. I stared into a computer long enough to check the database that tells me whether *cyberspace* ever has appeared in something I wrote. Sadly, yes, I did write it once. I'll try not to do it again.

I also wrote *doyenne* once. It means a woman who is the senior member of a group or the most experienced, as in Martha Stewart, the doyenne of domesticity-for-profit. Although doyenne pops up often in style-conscious magazines, real people never say it. I think I'll go the rest of my life without saying it, in part because I don't know how it's pronounced. I'd ask a maven, but I'm afraid of being squawked at.

—Tuesday, September 30, 1997

He's *copacetic thinking 'bout best words*

WE'VE HAD LISTS OF THE CENTURY'S BEST MOVIES, best books and best television shows. But what were the best words?

In the interests of starting an argument, I've made a list of the best words of the century.

I like most words, so it was really just a matter of finding some that didn't exist before 1900 and then eliminating anything that had to do with computers.

Computer people coin ugly terms such as *random access memory*. When they can't think of anything ugly enough, they steal good words such as *protocol* and *virtual* and ruin them. Still, I love computer people for inventing the hideously named word processor.

That aside, here are the Best Words of the Century:

Psychobabble — An incredibly useful word. It was coined around 1975, and I don't know how we managed without it. *Psychobabble* describes about three-fourths of everything spoken, written or sung in America in the 1990s. The self-help movement is almost entirely psychobabble; so are Celine Dion song lyrics, most Barbara Walters interviews and nine out of 10 business-management books.

Penicillin — How could a parent not love this word? Can you imagine parenthood without penicillin and the host of other "-cillins" that make our kids' infections easy to cure?

The word, born in 1928, comes from the penicillium fungus that produces penicillin.

Bubble gum — This is such a happy, airy term, and so familiar, but we've been saying it since only 1937. That's according to the Merriam-Webster Web site (www.m-w.com), a great place for word-lovers.

Other fun food words that only recently joined us: *jelly bean* (1905) and *cotton candy* (1926).

Beatnik — I love this word because a newspaper columnist coined it. The late Herb Caen of the *San Francisco Chronicle* was writing back when the beat movement was catching on in his town, Merriam-Webster says. Caen treated the movement with the proper irreverence. His April 2, 1958, column contained the first use of beatnik: "*Look* magazine, preparing a picture spread on S.F.'s beat generation (oh, no, not again!), hosted a party in a...house for 50 beatniks, and by the time word got around the sour grapevine, over 250 bearded cats and kits were on hand...."

We usually associate beatniks with the '50s and hippies with the '60s. But another great word Web site, Jesse's Word of the Day (www.randomhouse.com), says hippie started as '50s jazz slang for someone attempting to be hip. In the '60s, the sense changed to the sex-drugs-rock 'n' roll connotation.

Jesse is Jesse Sheidlower, senior editor in the Random House Reference Department.

Haywire — I like the agrarian roots of this word. Merriam-Webster says it comes from the light baling wire used to bind hay. Sometimes, cheap haywire was used to make slapdash repairs on defective machinery. From that, we arrived at *went haywire* to describe something that has broken down. Were we inventing such a term now, we'd say, "It went duct tape."

The haywire expression became common around 1920. We're so dependent on complex machines now that it feels good to use an uncomplicated term such as haywire when they fail.

Movie — It's a short, descriptive word for a modern pleasure, and it lacks the pretentiousness of film and cinema. *Movie* made its debut in 1912 as a shorter way of saying moving picture.

Speaking of entertainment, when HarperCollins publishing and *The Times of London* ran a contest to name the word of the century, the winner was *television*. I demand a recount.

Copacetic — Have you noticed that our slang words for expressing satisfaction tend to be satisfying in themselves? *Hunky-dory*, *dandy*, *cool* and *groovy* are all fun to say. So is copacetic. The Jesse's Word of the Day site says no one knows where *copacetic* originated. It may have roots in jazz or African-American culture. The earliest written use is from a 1919 novel.

I've run out of space without mentioning *jeep*, *bikini*, *nylon*, *neutron*, *robot*, *googol*, *laser* or about a thousand other 20th-century favorites. I'd list them all, but the word processor might go haywire.

—Saturday, June 19, 1999

Here's how to lose at Mega Millions like a real expert

OHIO HAS JOINED MEGA MILLIONS, a multistate lottery that offers bigger jackpots than the state's Super Lotto Plus.

Because the game is new, some residents are uncertain how to play. So To Speak presents this guide in an effort to aggravate the confusion:

• **Welcome.** The Ohio Lottery Commission welcomes you to the exciting new multistate game Mega Millions.

Now you can join residents of eight other states in the fun of buying colorful tickets that you will throw away in bitter disappointment two days later.

• **Odds of winning.** Your odds of winning the jackpot are 135 million-to-1, approximately the same as your chances of being struck by lightning while fending off an attack by a two-headed shark that was recently elected prime minister of Tajikistan.

In contrast, the odds of winning the state's Super Lotto Plus jackpot are 13.9 million-to-1 (same scenario, one-headed shark).

• **Two choices.** Option A: Buy Mega Millions tickets at any of the 9,100 Ohio Lottery retailers. Tickets cost $1 each.

Option B: Take a dollar from your wallet and throw it into the nearest storm sewer.

With either option, your odds are about the same.

• **Financial disclosure.** If you choose Option A, your costs for driving 5 miles to a retail outlet and buying one ticket will total approximately $2.60.

If you choose Option B, your costs for throwing a dollar into a storm sewer will total approximately $1.

In a year of weekly play, Option B participants will spend approximately $83.20 less than Option A participants.

The raccoons living in the sewer, meanwhile, will make $52.

The odds of a raccoon spending anything on a Mega Millions ticket are too small to calculate. Raccoons are not that dumb.

• **Strategy tips.** Select five numbers from 1 to 52 plus a sixth Mega Ball number from 1 to 52. These are your Mega Selections, otherwise defined as "losing numbers."

Use any method to pick the numbers. Some players choose based on dreams, street addresses or their children's birthdays.

Before deciding to have children, please consult a physician. The Ohio Lottery will not be responsible for children conceived in an effort to produce lottery numbers.

If you dislike the pressure of choosing, try the Auto Loss option, in which your losing numbers are selected randomly by computer.

If your six numbers match those in the Mega Millions drawing, pinch yourself. You are dreaming.

In the event you are not dreaming, redeem the winning ticket at an Ohio Lottery Commission office, then hold a news conference at which you announce you will not quit your job for a while.

Then quit your job.

• **Purpose of the game**. The mission of Mega Millions is to produce enough Mega Losers to ensure that the few Mega Winners do not drain all the profits.

Mega Millions is designed to entertain Ohioans while preserving the political careers of Gov. Bob Taft and the members of the General Assembly.

All Ohio profits go into an education fund that, when spread equally among the school districts, provides enough money to buy every child an extra chicken nugget in the cafeteria.

If you would like more information on Mega Millions, don't ask. You'll take all the fun out of it.

—Saturday, May 18, 2002

Prediction: U.S. relations will turn to (French) toast

NEWS BRIEFS FROM THE COLD WAR of the future between France and the United States:

France discharges diplomats

France expelled U.S. diplomats yesterday to protest an American decision to stage a bicycle race that competes with the Tour de France.

"This Tour de Cleveland, it is an insult to French honor," an official in Paris said. "And then the insensitive Americans tried to smooth things over by offering to let us cater the Iditarod. I spit on their arrogance."

The expulsion provided further evidence of deteriorating relations between the two former allies.

French bread crisis deepens

France yesterday rejected a U.S. demand that it stop making intercontinental ballistic baguettes capable of reaching American cities.

"For years, Americans have been attacking French cuisine with McDonald's, Spam and Wonder bread. We must defend ourselves," the French foreign minister said. "If that means carpet-bombing Peoria with a crusty sourdough, then so be it."

Cuisine-limitation talks broke off two months ago when the two sides couldn't reach an agreement on SpaghettiOs.

France had condemned the brand name, Franco-American, as part of a U.S. disinformation campaign seeking to blame the French for canned pasta.

U.S. denies Caribbean assault

French troops participating in war games claim they were confronted yesterday by U.S. troops on the island of Martinique.

The French, who have been staging military maneuvers as a show of strength, described the soldiers who forced them from a beach as "heavily armed."

U.S. diplomats denied that the confrontation involved the American military.

"It was a drill team from the University of Nebraska on spring break," an official said. "You know the French: They'll surrender to anybody."

Students protest 'delicacy'

American students on a goodwill mission to France were served the internal organs of a goose at a Parisian restaurant, the United States charged yesterday.

The incident shattered hopes that the cultural exchange might ease tensions between the two nations.

"They said it was 'fat-tay' or 'pat-tay' or something," said teacher Jerry Splerk, an American chaperon. "When I pressed them, they admitted it was goose liver."

Several students reportedly fainted.

"I mean, goose liver? I felt like I was on *Fear Factor*," said Ashley Callow, a sophomore at Abercrombie & Fitch High School in Bucyrus. "They said it's a delicacy. I'm like

'Yeah, right — we're dumb enough to fall for that.' "

Summit agreement reached

U.S. President Oprah Winfrey and French President Catherine Deneuve ended a three-day summit yesterday with a joint "statement of understanding."

The agreement commits both nations to immediate steps designed to lessen hostilities.

Highlights of the accord:

• The United States will ask TV stations to stop airing *Pepe LePew* reruns.

• France will apologize for the Renault.

• The United States will quit accusing France of entrapping western Europe behind "an iron curtain of haughtiness."

• France will no longer pretend that deconstructionism makes sense.

• The United States will encourage hotels to welcome French travelers with a "Mistresses stay free" offer.

—Friday, March 14, 2003

If he won as a write-in, things sure would change

TODAY, WITH SOME RELUCTANCE, I announce my candidacy for governor of Ohio.

Dick Brigode, a reader on the Far North Side, called recently to tell me he was making me a write-in candidate.

Although I had no intention of entering politics, I can't ignore the will of the people: Brigode has found a couple of voters who allow that I couldn't do any worse than Bob Taft or Tim Hagan.

Well, yes, I could — but let's not get into that right now. I have too much to do.

A candidate for governor needs a running mate and a first lady (my wife refuses to serve).

Brigode has agreed to be lieutenant governor, seeing as how the job requires only about two hours a month.

Anyone interested in becoming first lady (for ceremonial purposes only, my wife stresses) can send me a resume.

To respect anti-discrimination laws, I will consider male first ladies, although I'm a little uneasy about the idea.

As for my platform, I've fashioned a plan that represents a bold, new vision for Ohio. I developed it, I must emphasize, after long, lonely hours of solitary deliberation — so don't blame Brigode.

The details:

• Maurice Clarett stays at Ohio State for four years.

• No more uses of *gubernatorial.* I hate that word.

• Third-graders settle the school-funding controversy as a class project. I guarantee a solution fairer, clearer and more timely than whatever adults devise.

• A cellphone-transmission interlock prevents calls in a moving car.

• Quiet hours in Ohio are 11 p.m. to 6 a.m.

• The new Ohio Lottery slogan is: "If you're stupid enough to play, we'll be glad to take your money."

• No more congressional districts shaped like amoebas. I'm redrawing the districts as rectangles. I'll borrow some of West Virginia to straighten the edges.

• Upper Sandusky must stop confusing me with its location below regular Sandusky.

• TV weather reporters get only 30 seconds to present a forecast, and I'd better not see any sweeping arm motions. Just tell me when it will rain, for heaven's sake.

• If the sign says, "Right lane closed ahead / Merge left," then merge left! Immediately! Or you will be sent to prison!

• Telemarketers must use rotary phones, wear gloves as they dial and use no more than seven words. That ought to cut down on the calls.

• Kids: When your parents say no, they mean no. Stop that screaming.

• No more 1,000-page state budgets full of imaginary numbers and incomprehensible explanations. Instead, the budget consists of a giant thermometer outside the Statehouse.

• To keep from taking themselves too seriously, state lawmakers must wear fruit on their heads when in session.

• I–71 is renamed the Joe Freeway; and I–70, the Other Joe Freeway. A governor has to have a little fun.

—Friday, November 1, 2002

Ohio slycentennial moments: Concrete geese and Amish hippies make for an inspired past

WITH BICENTENNIAL FEVER SWEEPING THE STATE, we want to do our part to convey a sense of Ohio's rich past.

Our initial goal was to call attention to lesser-known facts about Ohio history.

Then we thought: Well, why stick to facts?

1763: The French and Indian War ends with Britain gaining control of the territory that would become Ohio. Alarmed by the spread of cynicism on the East Coast, the British issue the Proclamation of 1763, outlawing irony west of the Alleghenies.

1764: A little-known sect, the Suburbanites, sets out for the new territory in horse-drawn SUVs. Derided in New York City for keeping flocks of concrete geese and golden retrievers, they settle near present-day Westerville and slowly grow into a social force.

1800: John Chapman roams Ohio propagating apple trees, earning the nickname Johnny Appleseed. Meanwhile, the lesser-known Johnny Marzetti walks the state promoting his invention: a casserole of ground beef, tomatoes and noodles that leads to the development of school cafeterias.

1802: The Ohio Constitution is written. Among its lesser-known provisions: no putting on airs. Decades later, the provision is invoked to prevent the construction of the New Roman Coliseum and the Tower of London, Ohio.

March 1, 1803: Ohio becomes a state when the General Assembly meets for the first time in Chillicothe. The first governor, George Washington Courthouse, is the lone state employee.

March 2, 1803: The state payroll grows to 24,000 workers, all of whom have the day off.

1817: The first Ohio snowbirds head for Florida in covered wagons going 20 mph slower than the speed limit, turn signals blinking endlessly.

1824: The Panic of 1824 leaves Ohioans anxious and depressed. To lift spirits, Gov. Francis Dublin Coffman establishes the Comical Place Names Commission. Its efforts at humor live on in places such as Obetz, Kinnikinnick, Homer, Bobo and Pee Pee Creek.

1825: Ohio digs a series of canals. Skeptics call the canals folly, but they give rise to the canal historical-marker industry, which thrives to this day.

April 9, 1865: The Civil War ends when Robert E. Lee surrenders to Ulysses S. Grant. A half-hour later, the first Civil War re-enactors meet in a field near Bucyrus.

1877: An era that historians will call The Age of Obscure Presidents With Facial Hair begins with an Ohioan: Rutherford B. Hayes.

1902: Golf breaks out in Ohio. Scientists think the invasive recreation came from Scotland in the hold of a ship. During the next century, golf will transform Ohio's terrain, turning rich farmland and lush forests into bizarre expanses of unnaturally green lawn dotted with sand pits and men in loud shirts.

1903: The Ohio Centennial Exposition introduces the world to Buckeye innovations such as the potluck dinner, mulch and shyness. The exposition influences cultures worldwide: The Ottoman Empire, for example, holds a state fair.

1951: Woody Hayes, the new Ohio State University football coach, has his players invade Liechtenstein to show that they mean business.

1969: Hippies struggle to gain a foothold as Ohioans mistake the shaggy-haired youths for Amish. The hippies leave in frustration after being asked countless times whether they are selling apple butter.

1998: Culture wars flare:

• The teaching of biology is challenged by people who believe the stork brings babies.

• Vegetable-rights activists picket the Tomato Festival.

• Conservative legislators ban Eve Ensler's ground-breaking play until she produces a tamer Ohio version. Result: *The Medina Monologues*, in which blushing women from a northeastern Ohio town talk euphemistically about their bodies.

2003: Ohio lays plans for the tricentennial by building barns in all 88 counties. By 2103, they will look quaint enough to paint with the tricentennial logo.

—Friday, April 25, 2003
With Charlie Zimkus

When house starts Iraqin', don't bother knockin'

SADDAM HUSSEIN KNOCKED ON our door about midnight.

"I will stay here while my forces complete the annihilation of the American invaders," he announced.

And, before we could protest, he trooped in with a bunch of Republican Guards, Special Republican Guards and Extra Special Republican Guards.

Whether he was the real Saddam, I don't know.

He arrived with a dozen body doubles who spread themselves throughout the house.

Our kids gave them nicknames: "Living Room Saddam," "Kitchen Saddam," "Broom Closet Saddam."

"Saddam has chosen your house for a, uh, weekend getaway," the Iraqi information minister said. "He needs some time away from adoring crowds."

My family had a hard time adjusting to Saddam, who overstated everything.

"These egg yolks are too hard," he declared at breakfast. "Because of your failure, you will suffer the agonies of the damned."

The outbursts didn't impress me (I have bosses), but they had a big influence on the kids.

Asked to take out the trash, the sixth-grader lighted a cigar as she eyed me coldly.

"If you continue your arrogant demands, I will fling you into the fires of hell," she said.

His people insisted that I allow Saddam to sleep in my underground bunker.

"I don't have an underground bunker," I said.

"Certainly you do," said an aide, winking and subtly gesturing toward the basement.

"Oh, right," I said. "Yes, Saddam, I would be honored if you slept in my, um, bunker. Just spread out your sleeping bag next to the washing machine."

Instead, Saddam slept in the washing machine — for extra security.

He scared my wife half to death when she lifted the lid to run a load of underwear.

"I cannot live like this another minute," she told me. "When I got up this morning, the first thing I saw was Saddam shuffling down the hall in my bathrobe."

"I noticed," I said. "He pinned all his medals on it."

"I'm sick of this! Call the police."

I sheepishly told the police that Saddam was watching *Oprah Winfrey* in the family room.

"Right, and Josef Stalin is having tea with Attila the Hun over at my place," the desk sergeant said.

I called back minutes later to report that several dozen heavily armed Republican Guards were lounging in the dining room.

The police still wouldn't come.

I called again to say that the Fedayeen were setting up a defensive perimeter around the guest bath.

They still wouldn't come.

Finally, I told them I had spotted a zoning violation.

They came. (I live in the suburbs.)

Alas, the Iraqi information minister fooled them: He described the gathering as a convention of Saddam impersonators.

"Saddam impersonators?" a cop asked.

"Yes, harmless entertainers," the minister said — "like Elvis impersonators, you see."

As if on cue, the Saddams broke into a chorus of *Love Me Tender*.

They're as wily as advertised.

Just as I was becoming resigned to a long stay, the minister announced that Saddam was departing.

His people packed up everything — the inflatable Saddam statues, fold-up Saddam murals, portable Saddam shrines — and left in a hurry.

"Saddam is a man of travel," the information minister said, "especially in recent days."

—Monday, April 14, 2003

Same-sex unions attractive to heterosexuals

SO TO SPEAK IS VENTURING INTO the future to see how the controversy over same-sex marriage turned out.

Here's a report from 2006:

Same-sex marriage, legalized in December 2004 by the U.S. Supreme Court's decision in O'Donnell v. Ashcroft, has changed the face of the nation.

As expected, thousands of gay couples wed in the months after the ruling. Protests diminished as people grew accustomed to seeing two little grooms atop the wedding cake.

What no one foresaw was that same-sex marriage would catch on with heterosexuals, too.

Men, weary of trying to please women, married their drinking buddies. Women, fed up with uncommunicative men, wed their confidantes.

Typical of the heterosexual marriage partners are Lenny and George of Columbus. Both 23, they have been married a year and say they are a contented couple, though with separate bedrooms.

"It's just easier to be married to a guy," George said. "Women are always bugging you to talk to them. All I have to do for Lenny is grunt at him a couple of times a day. Plus, I'm covered by his health insurance."

They met in 2005, when both were working at a warehouse and struggling to make ends meet. One night, while having a few beers after a softball game, one of them facetiously suggested that they should marry.

"It started as a joke," George said. "But then we began looking at the advantages: We could register at Ace Hardware for wedding gifts. We could get cheaper car insurance. And whenever a girlfriend started bugging me to make a commitment, I could say, 'Sorry, already married.' "

Today, the couple lives in a comfortable house, accented with beer-can pyramids in the windows and a motorcycle in the dining room. (Lenny is rebuilding the carburetor.)

Asked to assess the state of their relationship, George shrugged.

"OK, I guess," he said. "We've actually never talked about it."

Lenny, the more emotionally expressive of the pair, admitted they'd had to work out some issues in the early months.

"We'd argue about the designated hitter or whether Shaquille O'Neal is overrated, and then I'd go off and brood for hours. Things got better when he started validating my sports feelings."

As for their love lives, George is dating Jennifer, who is wed to Allison.

Jennifer said her same-sex marriage and opposite-sex romance have many advantages.

"When you come home from a date, it's nice to have a spouse to tell all about it."

Jennifer, 24, married Allison after divorcing Frank, a macho type.

"The first thing Allison and I did after the wedding was take a long car trip just so we could ask for directions. It was so liberating."

Jennifer said the most stressful part of the relationship came when she and Allison had to share the spotlight at the wedding.

"In retrospect, I suppose two garter-throwing ceremonies was overkill."

Like George and Lenny, they sleep in separate bedrooms in their four-bedroom house (the two others are for shoes and teddy bears).

George and Lenny, meanwhile, are planning a belated first-anniversary celebration (both forgot) at a Poconos resort.

"We're staying in the honeymoon suite," Lenny said dreamily. "It has a football-shaped bathtub, a turkey fryer and a big-screen TV with his-and-his remote controls."

—Friday, March 12, 2004

Six

The Natural World

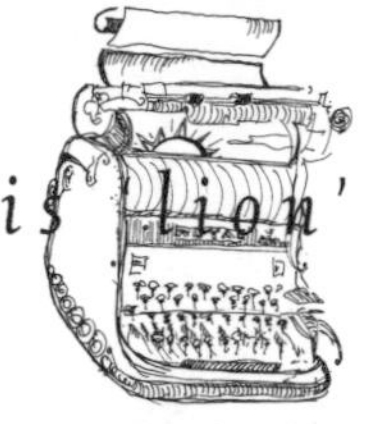

Mother Nature is lion' down on the job

I LIKE NATURE, THE UNIVERSE, evolution and all that.

I make that statement upfront so that the scientific crowd doesn't deluge me with angry letters again.

When I criticized Pluto (the planet, not the cartoon dog), all hell broke loose.

I merely suggested that its elliptical orbit signaled a certain sloppiness that should be corrected. Round it off, Pluto — that's all I said.

Astronomers called me ignorant and out of touch with reality. Sheesh!

So it is with hesitation that I present a list of things in nature that aren't fair.

Some people will accuse me of another scathing attack on the universe. Yet I cannot keep silent in the face of injustice.

Here's my list of gripes with the natural world:

• **Carbon monoxide is colorless and odorless.**

Come on, that's just not right. Even rattlesnakes have the courtesy to provide a warning before poisoning someone.

At a minimum, carbon monoxide ought to be a sickly yellow and smell like spoiled egg salad.

The gas has an unfair advantage, and I know I speak for many when I say: Get an aroma.

• **Heart attacks don't hurt enough.**

Can we get some sanity in the pain world, please?

Stub your toe, and it hurts so badly you want to die. Have a heart attack, and, yes, you might feel terrible pain. Then again, you might just feel as if the lunchtime burrito didn't go down right.

Meanwhile, your left ventricle is seizing up.

That's ridiculous: Pain should be on a graduated scale, corresponding to the seriousness of the cause. A tumor ought to hurt more than a toothache.

Is that asking too much?

• Male lions don't do anything.

I can't begin to count how many *Animal Planet* episodes have been ruined for me by the spectacle of good-for-nothing male lions.

The females spend all day caring for the cubs. Then, at dinnertime, they bring down a 1,500-pound Cape buffalo with a head as wide as a pickup truck.

And the males?

They lie around all day, looking regal, while the females are confronting the slashing horns of death. Only after the buffalo is safely dead do they trot up to shoo the others off the carcass and eat all the good parts.

I hate that.

• Days start getting shorter at the beginning of summer.

Oh, sure: Spoil the party when it has just begun.

Summer stirs expectations like no other season of the year. No sooner does it arrive than the sun starts knocking off early, darkening the mood.

Hellooo! The longest day of the year should occur at the end of July.

We've had our fill of sunburns and humidity by then, so we're ready to contemplate the start of fall. Timing is everything.

• All primates — except humans — have prehensile feet.

Nature gave nimble, grasping feet to gorillas, chimpanzees and orangutans but skipped the primate that could really use them.

Is that fair? Do gorillas drive cars, play pianos or hoist beer mugs? When was the last time a chimp tried to eat a hamburger while piloting a 4,000-pound SUV at 75 mph?

Are you listening, nature?

—Monday, September 23, 2002

COSI *celebrity has been around, but she's no fossil*

SHE'S TALL, THIN AND EXOTIC, WITH AN UNTAMED air that seems to enhance her allure.

Sue, the tyrannosaurus skeleton, has come to Columbus for a summerlong engagement at COSI. Until arriving here, she had never granted a media interview. But recently she sat down with So To Speak for a candid and exclusive conversation.

Is the creature who bares her ribs as enthusiastic about baring her soul? Let's find out:

Q: Sue, there's been lots of gossip about your life as a hell-raiser in the Cretaceous period. What can you tell us?

A: When I was young and carnivorous, I was young and carnivorous. That's all I'm going to say.

Q: Are we being a bit evasive?

A: Look, it was an interesting time in my life. A time of challenge and personal growth — I gained 1,100 pounds in one year and had to disembowel a rival female, for example.

But it was 65 million years ago. I'm not who I was back then. Why is that so hard for people to understand? I mean if Jesse Ventura can become respectable, surely a tyrannosaurus can.

Q: Let's change the subject. How's the COSI show going?

A: Wonderfully. Everyone has been so supportive. It's very exciting to be on display here.

Q: I'm amazed that you can stand in one pose for hours at a time. How do you do it?

A: Yoga. I'm actually doing yoga the whole time. Of course, if it were up to me, I'd be in the lotus position, but the COSI people thought that would be too distracting. They said kids are coming to see a nightmarish predator, not the Dalai Lama.

Q: What's the nicest thing that's happened to you here?

A: Oh, this was so sweet. It was opening day. Kids everywhere. They're trying to climb my vertebrae, throwing paper airplanes into my eye sockets, screaming. I was sooo tired.

But when I got back to my room that night, I discovered that some thoughtful fans had arranged for the hotel to leave 400 pounds of raw meat on my pillow. I wasn't even hungry, but that one little gesture really touched me.

Q: Speaking of food, Sue, I'm sure you've heard the speculation that you're anorexic.

A: Yes. And let me say right now, it is completely unfounded.

If you're a female celebrity and you're a little bony, the tabloids are going to spread rumors that you have anorexia. Calista's been through it; Lara Flynn Boyle's been through it. And now me. It's so unfair.

Q: So, if we looked in your refrigerator, what would we find?

A: (Laughs). Just regular stuff. Bottled water, some fruit, maybe the partially eaten carcass of a large mammal. I don't do big meals, I nibble.

Q: Are you looking forward to the new *Jurassic Park* movie?

A: Absolutely not.

I was really upset by the first one. The way that movie portrayed the tyrannosaurus community! Killing defenseless goats, thrashing around amid palm trees, selfishly seeking to fulfill primitive needs. What do they think I am, a *Survivor* contestant?

Q: What question are you tired of answering?

A: (Sighs). I get really tired of all the curiosity about my arms. You know: Why are her arms so short? How did she hold prey? Does it mean she was just a scavenger? Blah, blah blah.

Hellooo! I didn't need long arms. I was taking down herbivores, not boxing Rocky Marciano. How hard do you think it is to knock off a stupid apatosaurus anyway? Half the time, all I had to do was jump out from behind a tree and say boo. Sheesh!

Q: You've never been known as a political animal, but recently you've opposed President Bush's energy plan. Why?

A: The emphasis on fossil fuels is very offensive to me. Every time I fill the tank, I have to wonder: Is this Uncle Fred I'm pumping into the Chrysler? It's uncivilized.

Q: What do you think of Columbus?

A: I absolutely love it. What a great town. The omnivores are all so plump and slow-moving.... Oh, dear, did I say omnivores? I meant the people. (Flashes a devilish grin.) Well, you know what they say: You can take the girl out of the Mesozoic Era, but you can't take the Mesozoic Era out of the girl!

—Tuesday, May 29, 2001

Dogs' agenda may make you paws and reflect

NATIONAL DOG WEEK BEGINS SUNDAY. You can look it up.

To mark the occasion, I interviewed the president of the National Association of Dogs. He's a collie-retriever mix from Chicago who wants dogs to be more active in public affairs.

Here's how it went:

Q:What's the mission of the National Association of Dogs?

A: To raise awareness of dogs, improve our image, and get our issues before the public. We're just like any other special-interest group.

Q: Aren't there some key differences, though? For example, most special-interest groups don't drink out of the toilet.

A: I'm glad you brought that up. Drinking out of the toilet, as you put it, is actually a water-conservation measure that we're justifiably proud of. Dogs have long been at the forefront of protecting the environment, but we've received little credit for it.

Q: How else have dogs protected the environment?

A: We've done so through a host of measures. For example, our trash-eating initiative has been extremely successful. Statistics show that it has kept 10 million tons of waste out of landfills in the past decade alone.

Q: So when I see a dog turning over my trash cans...

A: You should thank him. Maybe even set out the dinner dishes for him to lick, as a way of showing appreciation.

Q: You say you want to raise awareness. But aren't people already aware of dogs? I mean, they bark.

A: We prefer to call it "expressive vocalization." And it's a misconception that dogs do too much of it. Our polling proves that.

We asked 1,500 randomly selected dogs how much they bark, and the results showed that barking is down a full 1 percent since 1998. In particular, barking at the moon, the wind, and pieces of paper blowing across the yard have fallen off.

Q: That's hard to believe.

A: We use some of the same polling companies as the Republican and Democratic parties do.

Q: No wonder it's hard to believe. You mentioned issues. What are some issues that dogs are concerned about?

A: Fire safety is a big concern. We're lobbying Congress to approve a $2 billion appropriation to put 100,000 more fire hydrants on the streets.

We've also proposed a $10 billion initiative to pay for more effective controls on the exploding cat population. We think there are too many of them on the streets. In homes, too, for that matter.

Transportation issues also concern us. Did you know that every year, 20,000 dogs are injured trying to bite the tires of moving cars? We think they should be made of a more chewable material.

Q: What about other issues? Health care? Gun control?

A: We don't care about gun control. We would like to see something done about that pepper spray the mail carriers use.

The rising cost of health care also is a big issue.

In fact, there's a consensus in the dog community that far too many trips are being made to veterinarians' offices. Too many shots, too many teeth-cleanings, too many examinations on cold steel tables. And don't even get me started on neutering.

We'd be satisfied to limit doctor visits to only the most serious cases, such as when we eat a 5-gallon tub of discarded restaurant grease or are disemboweled by a cornered raccoon.

Q: Will the National Association of Dogs be endorsing a candidate for president?

A: We're undecided at this point. The Gore alpha-male strategy was an obvious attempt to woo dogs, but we found it unconvincing. A true alpha male would have gone to the Democratic National Convention, peed in every corner to mark his territory, and sired a few litters before he left town.

As for Bush, well, he seems like an amiable guy, yet his struggles with the language alarm us. He went to Yale? I know graduates of Canine Training College who have better grammar.

Q: How would you like to see National Dog Week observed in America?

A: I'd just like to encourage everyone to enjoy dog traditions. Go out and sniff a friend. Chase a few squirrels up trees with the kids. Find something dead to roll in.

Q: Should we drag our butts across carpet, too?

A: Why not? It's a celebration.

—Tuesday, September 19, 2000

Groundhog refuses to chuck tradition

HE'S THE RODENT WHO BECOMES a rock star every Groundhog Day.

But has Punxsutawney Phil also developed a rock star's lifestyle?

That's been the buzz in Punxsutawney, Pa., where the celebrity woodchuck is making his famous weather prediction today.

Rumors of unreasonable demands, rowdy behavior and an inflated ego have been following Phil for months.

The rumors range from the mild (Phil's contract specifies organic carrots in his dressing room) to the wild (he got drunk at a party thrown by Martha Stewart and ate her perennial border).

Recently, So To Speak caught up with Phil for an interview in his comfortable but modest winter digs. He summers on Cape Cod and has a home in Beverly Hills but claims he will always consider the Punxsutawney burrow home.

Dressed in a warmup suit, Phil looked surprisingly fit and alert for a rodent known for packing on pounds, then sleeping away the winter.

"Credit my personal trainer," Phil said as he nibbled on a salad. "I always used hibernation as an excuse not to exercise. She convinced me that being semi-comatose with a body temperature of 39 degrees is just another obstacle to overcome."

Phil was evasive about rumors, reported first in *People* magazine last fall, that he has undergone liposuction. His handlers were reportedly appalled.

"On Groundhog Day, the public expects a plump, sleepy groundhog, not an aerobics instructor in a fur coat," one handler said. "I'm all for physical fitness, but let's have some respect for tradition here."

The groundhog is offering no apologies, either for his diminishing waistline or his expanding ambition.

He has signed a licensing deal with Nike (GroundTogs furry athletic wear), plans to direct a biopic of his early years in a rough burrow (working title: *'Hoodchuck*) and says he might do an album of rodent tunes (*Muskrat Love, The Chipmunk Song, Three Blind Mice*).

"I need to stretch," Phil said. "I don't want to go to my grave being known just for crawling out of a hole. I don't want to be the Saddam Hussein of groundhogs."

Phil's determination to cast a bigger shadow is reflected in his entourage. It includes a makeup artist, fur stylist, personal secretary, publicist and burrowing crew.

He also has been keeping a team of lawyers busy with his lawsuit against Dr. Phil.

Phil alleges that the TV psychologist copied his name, appearance and act.

"Pretty obvious, isn't it?" Phil said. "I have whiskers; Dr. Phil has a mustache. People look to me for knowledge; people look to him for knowledge. I originated the persona of the chubby know-it-all, and he's trying to steal it."

As for the tales of a woodchuck run amok, Phil says the stories are exaggerated. He acknowledges hitting the party circuit ("always with a designated digger") but says he remains a very grounded groundhog who holds Feb. 2 sacred.

"A few years back, I woke up on Groundhog Day with the worst hangover of my life. I wanted to say I saw my shadow just so I could go back to bed. But I resisted the temptation. I have too much respect for the integrity of the process."

Suddenly, a yawn overtook him, and Phil announced that the interview was over.

"Louis," he called to his personal assistant, "prepare my bed. I need to get into character."

—Monday, February 2, 2004

They're cockroaches to us; to him, they're research subjects

CLEVELAND — Professor Roy Ritzmann stuck his hand into a bucketful of cockroaches and withdrew a healthy-looking specimen.

"Blaberous discoidalis," he said. "The death-head cockroach."

It gets its nickname from the skull-shaped marking on its back, a colorful splotch that would be pretty if it weren't on a cockroach. I decided the bug was best admired from a distance of at least 3 feet.

We were standing in a small, dimly lighted storeroom at Case Western Reserve University — just me, Ritzmann and his research subjects, several thousand male cockroaches in about a dozen covered buckets. This is one room where cleaning up a spill would require more than paper towels.

Ritzmann keeps the storeroom locked, although I can't imagine that theft is much of a problem.

"We try to work mostly with males," Ritzmann said. "If a female gets loose, she can drop an egg case."

I went to see Ritzmann, 51, because he and his colleagues study a strange subject — cockroach locomotion. This confirms my long-held belief that if you can conceive of a subject, you can find a professor somewhere who studies it.

That's not to suggest that Ritzmann's studies are trifles. You would be surprised what a knowledge of cockroach locomotion can be applied to.

Ritzmann, a native of Chicago with a doctorate in biology from the University of Virginia, has spent 25 years studying how the cockroach moves. He uses death-head cockroaches and their cousins, the smaller, faster American roaches. As research subjects go, roaches are cheap, easy to obtain and possessed of agility and speed that humans only can dream of.

Ritzmann's office and lab are in a nest of cramped rooms in the Biology Building at Case Western. His desk is decorated with a ceramic mixing bowl crawling with fake cockroaches. Anyone bothered by that better venture no farther.

Look in on Ritzmann's lab at the right time, and you might see a researcher laboriously implanting hair-thin electrodes into a live cockroach's legs. Or you might see a roach on a treadmill, a roach climbing a tiny ladder or a roach pinned to a greased tray, running in place, while a video camera records its movements.

Ritzmann, who specializes in behavioral neurobiology, has a particular interest in the insect's "escape response." That's a scientist's term for when the bug's nervous system tells it to run like hell.

To simplify a complicated story, the insects have tiny hairs on their back ends that can detect a puff of wind, such as that generated when a hungry predator lunges or a human foot descends with murderous intent.

These hairs send messages to the legs, which turn the cockroach away from the direction of the wind and send it scurrying in a fraction of a second.

Roaches have other unnerving abilities, too. Ritzmann told me that one can disconnect its brain from the rest of its nervous system and it still will eat and walk.

Ritzmann's research has focused on mapping the nervous-system responses and leg action that make things like this possible. Loathsome as they are, cockroaches may have something to teach us.

A robot that could maneuver over and around obstacles as effortlessly as a cockroach would be valuable, which is why the work of Ritzmann and some Case Western colleagues has caught the attention of the U.S. Navy. It wants a robot that could clear land mines, and the Case Western team has built some steel prototypes that pattern their movements after the insect. That's right — it's cockroach engineering.

"From a robot standpoint, (insects) have solved just about every mechanical problem in the world," Ritzmann said.

The Case team is working on its third prototype, a machine that looks eerily like the creature that inspired it. Ritzmann said that when the robot was demonstrated at a conference, a participant with an aversion to insects said, "You realize, this is my worst nightmare — a 30-pound steel cockroach."

Ritzmann himself doesn't have an unequivocal love of his research subjects. They're fine in the laboratory, he said. But if he found them at home?

"I'd call the exterminator, like anyone else. I'm not nuts."

—Thursday, March 18, 1999

Mellow yellow: In moderate praise of the tenacious, edible dandelion

IF YOU'RE GOING TO HAVE A LAWN IN THIS COUNTRY, you have to develop a dandelion philosophy. Because dandelions happen to all of us.

I'm a dandelion moderate myself. I like some in the lawn to create a festive atmosphere, keep life in perspective and remind me of my Italian grandmother.

When spring came, my grandmother harvested food from the lawn. She went out and dug the tender young dandelions out of her yard, our yard and anyone else's yard available.

She lived just a few blocks away from us, and in early April it was not unusual to glance out the back door and spot her there in the grass, unannounced, stooping over her work. She heralded spring, like a robin in a head scarf.

This was at a time when people were less prone to carpet-bomb their yards with chemicals to exterminate dandelions.

People who do that are militant anti-dandelion conservatives. They've taken old-style anticommunism and transferred it to lawns. They'll tell you that dandelions will spread and take over the property if you don't control them. Please. We're talking about a salad green here, not a rebel insurgency.

On the other hand, I can't align myself with modern dandelion liberals either. A lot of them are herbalists or aromatherapists or other new-agey types who claim that dandelions are powerful storehouses of vitamins that will enable us to do yoga and picket for animal rights into our 90s.

I'm suspicious of people who ascribe magical qualities to food.

My grandmother, from the last generation with any sense about food, needed no rationale for picking dandelions other than that they were edible. Members of her generation believed eating was healthy in general. Their children, sadly, took this to the extreme of embracing TV dinners.

The pendulum then overswung in the opposite direction, and today we have assorted food cultists who worship organic vegetables, swear by low-carb diets, bathe in soy or eat nothing but Subway sandwiches.

Dandelion liberals also won't tell you the truth about dandelions, which is that they basically taste like hell. They come in two flavors: bitter and real bitter.

I don't mind because I like an intense flavor experience. Still, if you want your offspring to adopt your dandelion philosophy, it helps to be honest with them. Dandelions can be disillusioning to the young, who bite into everything hoping for the taste of brownies.

Now let's address the larger spiritual, social and aesthetic implications of dandelions.

First, they are such tenacious survivors, I have to conclude that God created them to humble us. A plant capable of putting down a 6-inch taproot in a sidewalk crack has character humans can only dream of. I suspect some people kill dandelions to avoid feelings of inadequacy.

Second, I believe that children who grow up around dandelions have a better understanding of yellow. I mean the yellow of a warm day in May, when the dandelions are blooming and the sun feels so good you can't believe it causes melanoma.

Speaking of children, I don't see how they can learn the joy of giving if they never have a chance to present their mothers with the classic preschooler bouquet: three wilted dandelion blossoms in a glass of water.

Kids aside, there's nothing more forbidding than a stretch of chemically enhanced, hypergreen, dandelionless lawn. It's the outdoor equivalent of white carpet. You shouldn't feel like you have to take off your shoes to walk on someone's grass.

Therefore, I always maintain a few dandelions to put visitors at ease.

They think: Oh, here's someone secure enough to let a few dandelions show. That, and the rust spots on his car, tell me he's not trying to overcompensate for shortcomings in other areas. At the same time, we know he's not a drug-addled libertine by the fact that his broadleaf weeds, while noticeable, fall well short of a code violation.

Granted, I could be reading too much into it.

You, of course, are free to develop your own dandelion philosophy, although you'd be a fool not to adopt mine. In any case, develop one quick. The dandelions are already sprouting.

—Tuesday, April 3, 2001

About The Author

JOE BLUNDO'S COLUMN, SO TO SPEAK, began appearing in the lifestyle section of *The Columbus Dispatch* in May 1997. It's a mix of humor, human interest and information.

In 2002, Joe won the National Society of Newspaper Columnists contest for humor writing in large newspapers.

He has been at *The Columbus Dispatch* since 1978. He started as the night police reporter and also has worked as a suburban reporter, a copy editor, an assistant Metro editor, Accent editor and a Home & Garden reporter.

He was born in New Castle, Pa., on April 17, 1954, to Catherine and Ben Blundo. An avid reader as a child, he became interested in writing while a student at New Castle High School.

After graduation, he enrolled at Westminster College in New Wilmington, Pa. He left there in 1973 to study journalism at Kent State University. He received his bachelor's degree from Kent in 1975, the same year he married Deborah Robinson of Westerville, Ohio.

Joe started his career at *The Parkersburg Sentinel*, where he worked from 1976 to 1978 as a reporter.

He and Deborah have two children and live in Worthington, Ohio.